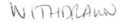

Atlas Florae Europaeae

I

ATLAS FLORAE EUROPAEAE
DISTRIBUTION OF VASCULAR PLANTS IN EUROPE

I

1
Pteridophyta
(Psilotaceae to Azollaceae)

2
Gymnospermae
(Pinaceae to Ephedraceae)

Edited by
Jaakko Jalas & Juha Suominen
on the basis of team-work by European Botanists

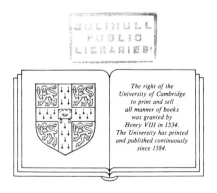

The right of the
University of Cambridge
to print and sell
all manner of books
was granted by
Henry VIII in 1534.
The University has printed
and published continuously
since 1584.

CAMBRIDGE UNIVERSITY PRESS
Cambridge
New York   New Rochelle   Melbourne   Sydney

Published by the Press Syndicate of the University of Cambridge
The Pitt Building, Trumpington Street, Cambridge CB2 1RP
32 East 57th Street, New York, NY 10022, USA
10 Stamford Road, Oakleigh, Melbourne 3166, Australia

First published in separate volumes by
The Committee for Mapping the Flora of Europe and Societas Biologica Fennica Vanamo, Helsinki
Volume 1 1972
Volume 2 1973

First published as a compendium by Cambridge University Press 1988

Printed in Great Britain at the University Press, Cambridge

British Library cataloguing in publication data

Atlas florae Europaeae: distribution of Vascular plants in Europe
1. Plants – Europe – Maps
I. Jalas, Jaakko   II. Suominen, Jaha
912'.158194   QK281

Library of Congress cataloguing in publication data

Atlas florae Europaeae: distribution of vascular plants in Europe / edited by Jaakko Jalas & Juha Suominen.
p.   cm.
Contents: I. 1. Pteridophyta (Psilotaceae to Azollaceae). 2. Gymnospermae (Pinaceae to Ephedraceae)
– II. 3. Salicaceae to Balanophoraceae. 4. Polygonaceae. 5. Chenopodiaceae to Basellaceae.
– III. 6. Caryophyllaceae (Alsinoideae and Paronychicideae).
7. Caryophyllaceae (Silenoaideae)
ISBN 0 521 34270 8 (v. 1)
1. Phytogeography – Europe – Maps – Collected works. I. Jalas, Jaakko. II. Suominen, Juha.
QK281.A85 1988
581.94 – dc19   87 – 33834

ISBN 0 521 34270 8

GENERAL PREFACE

I am very pleased to be given the opportunity to help in introducing to a wider public these distribution maps, to which I have made such frequent and profitable reference over the past few years – all the more so since it enables me to justify to some small extent my nomination as a member of the advisory committee for the *Atlas*, of which I have hitherto been a rather inactive member.

Towards the end of 1961 I became conscious, in the course of my editorial work on the first volume of *Flora Europaea*, that in many contributions which were satisfactory taxonomically the data on geographical distribution were defective or inaccurate. Sometimes countries were omitted from mere inadvertence; sometimes they were given for a plant which had never been more than a casual and was long since extinct in the field but still flourishing in the Florae; very often the mistakes arose from a failure to take account of territorial changes in the period 1914–1945. It was astonishing how many experienced botanists seemed to imagine that the Carpathians were still in Hungary, or that Finland still had an arctic coast. To remedy this state of affairs I assumed, with the consent of the editorial committee, responsibility for revising geographical data in all accounts. But time was short, and when the first volume went to press I had been able to correct only some of the grosser errors. In the remaining volumes I believe the geographical data to be be more accurate.

The maps so far published in *Atlas Florae Europaeae* all relate to plants which are described in the first volume of *Florae Europaea*, and they are of especial value in serving to correct the rather numerous errors in the latter. They have been of enormous help to me in the preparation of the second edition of this volume. But their usefulness goes far beyond that. In *Flora Europaea* considerations of space limited us to a phrase of about 25 words at most to sum up the distribution. It is in most cases reasonably accurate, but the *Atlas*, besides correcting some minor inaccuracies, fills in a lot of detail that could not be included in the printed phrase. For *Stellaria holostea*, for example (I choose a species which is free from any taxonomic doubts or difficulties), *Flora Europaea* rightly says that it is 'rare in the Mediterranean region', but the *Atlas* shows how unevenly distributed is this rarity: the plant is completely absent from southern Spain, extremely rare in Greece and Albania, but relatively frequent in Italy. Or again, take *Spergularia salina* (= *S. marina*), for which *Flora Europaea* gives 'Coasts of Europe and inland saline areas'. Few readers can have realized how extensive these inland saline areas are until the *Atlas* revealed them.

The *Atlas*, of course, is not perfect; its editors would be the last to claim that it is. But most of its shortcomings arise from the fact that the editors are largely dependent on the reports they receive from the various countries of Europe; and although they have built up an impressive team of collaborators (far more complete than that achieved by the editors of *Flora Europaea*), it is inevitable that they are sometimes supplied with questionable or defective data, or, worse still, with none at all. So we find occasional species which are native throughout Spain but alien throughout Portugal; species which are present in every square in Bulgaria but curiously rare in all the neighbouring countries; and numerous species for which the sparse distribution in the more remote parts of the USSR probably tells us more about the distribution of Russian botanists than of plants.

More disconcerting is the occasional failure of some regions to supply any data at all; in recent issues there have been alarming blanks in Romania and Ukraine for several species. In these circumstances dots for relatively rare species can be provided, at least in part, from the herbaria and literature, but for very common species this is impossible. It is difficult to know what the editors can do in such a situation; my only suggestion is that they should indicate rather more clearly than at present (even at the sacrifice of a little tact) which of the blanks indicate absence of plants and which of them absence of data.

I mention these limitations of the maps not for the sake of carping, but to enable the reader by bearing in mind these difficulties to interpret the maps with more discrimination, and also because these limitations have been dealt with at greater length in a recent article by Dr Holub (Norrlinia 2: 107–115 (1984)). Although it contains some judicious praise of the *Atlas*, it does not hesitate to suggest a large number of ways in which it might be improved. The breaking down of all aggregates into their segregates (according to whose taxonomy?); the treatment of apomictic genera on a heroic scale (with maps of 'at least 200 or 300 species of *Rubus*'); the modification of the grid system in certain areas; the ironing out of differing views on native, alien or casual status; greater detail in data on recent increase or decline; fuller citations from literature; the immediate preparation (out of sequence) of maps for the most critical genera – these are some of the improvements which he suggests. They are all desirable, but they would need the services of about a hundred more floristic botanists, at least a dozen extra critical taxonomists, and at least one trained diplomat with a generous travelling allowance. Holub, one suspects, wishes, as did Omar Khayyam, that he could 'with Fate conspire, to take this sorry scheme of things entire . . . and then remould it nearer to his heart's desire'.

Pending this consummation, let us be thankful for what we have got, and realize that in this imperfect world we are not likely to get much more, and let us trust the editors to make the best decision when faced with the many awkward problems which beset them.

The thanks of all European botanists (and of many outside Europe) are due to the editors of the *Atlas* for their enterprise in initiating the project and for their tenacity in continuing it. Thanks are also due to the government of Finland for the generous and far-sighted support which it has provided. It gives me special pleasure as an Irishman to see a country not so very much more populous than my own, and nearly as isolated on the periphery of Europe, playing such an important role in forwarding the scholarship of Europe as a whole.

June 1987

D. A. Webb
Trinity College, Dublin

1
PTERIDOPHYTA
(PSILOTACEAE TO AZOLLACEAE)

CONTENTS

Index to Vol.1 is at the end of Vol.2
Base map is at the end of Vol.2

THE COMMITTEE FOR MAPPING THE FLORA OF EUROPE

SECRETARIAT

J. JALAS, Helsinki (Chairman of the Committee)
J. SUOMINEN, Helsinki (Secretary General)

ADVISERS

T. W. BÖCHER, København
A. R. CLAPHAM, Sheffield
E. HULTÉN, Stockholm
D. A. WEBB, Dublin

COMMITTEE MEMBERS ACTING AS REGIONAL COLLABORATORS

Albania (Al)	I. MITRUSHI, Tiranë	
Austria (Au)	F. EHRENDORFER, Wien	H. NIKLFELD, Wien
Belgium (Be, excl. Luxembourg)	E. VAN ROMPAEY, Antwerpen Assisted by: L. DELVOSALLE, Bruxelles J. E. DE LANGHE, Antwerpen	A. LAWALRÉE, Bruxelles
British Isles (Br, Hb, Channel Islands)	F. H. PERRING, Abbots Ripton Assisted by: M. N. HAMILTON, Abbots Ripton	
Bulgaria (Bu)	S. KOŽUHAROV, Sofia	
Czechoslovakia (Cz)	J. FUTÁK, Bratislava	J. HOLUB, Průhonice
Denmark (Da, Fa)	A. HANSEN, København	
Finland (Fe)	J. SUOMINEN, Helsinki Assisted by: T. ULVINEN, Oulu	
France (Co, Ga)	P. DUPONT, Nantes Maps checked by several French botanists	

German Democratic	H. Meusel, Halle	E. Weinert, Halle
Republic (Ge: DDR)	Assisted by:	
	W. Fischer, Potsdam	H.-D. Krausch, Potsdam
	F. Fukarek, Greifswald	W. Müller-Stoll, Potsdam
	W. Hempel, Dresden	J. Pötsch, Potsdam

German Federal H. Ellenberg, Göttingen U. Hamann, Bochum (—1966)
Republic (Ge: BRD) H. Haeupler, Göttingen
Assisted by:
P. Schönfelder, Stuttgart, and collaborators in the mapping of the Central
European Flora

Greece (Cr, Gr) P. Critopoulos, Athínai
Assisted by:
W. Greuter, Genève (Kriti) H. Runemark, Lund (Aegean Islands)
G. Lavrentiades, Thessaloniki

Hungary (Hu) Z. E. Kárpáti, Budapest R. v. Soó, Budapest
Assisted by:
A. Terpó, Budapest

Iceland (Is) E. Einarsson, Reykjavík
Assisted by:
H. Björnsson, Öraefi B. Jóhannsson, Reykjavík
I. Davídsson, Reykjavik I. Óskarsson, Reykjavík
H. Guttormsson, Neskaupstadur S. Steindórsson, Akureyri
H. Hallgrímsson, Akureyri T. Thorsteinsson, Reykjavík

Italy (It, Sa, Si) E. Nardi, Firenze J. Pignatti, Trieste (—1966)
G. Moggi, Firenze

Jugoslavia (Ju) E. Mayer, Ljubljana

Luxembourg L. Reichling, Luxembourg
(Be, excl. Belgium)

Netherlands (Ho) S. J. van Oostroom, Leiden

Norway (No, Sb) K. Faegri, Bergen

Poland (Po) J. Kornaś, Kraków
Assisted by:
D. Fijałkowski, Lublin L. Olesiński, Olsztyn
A. Frey, Kraków A. Pacyna, Kraków
J. Guzik, Kraków A. Skirgiełło, Warszawa
J. Jasnowska, Szczecin A. Sokołowski, Białowieża
K. Kępczyński, Lublin H. Rutowicz, Łódź
M. Kopij, Warszawa A. Zając, Kraków
J. Mądalski, Wrocław E. U. Zając, Kraków
J. Mowszowicz, Łódź W. Żukowski, Poznań
R. Olaczek, Łódź

Portugal (Az, Lu)	J. do Amaral Franco, Lisboa (1967—)	A. R. Pinto da Silva, Oeiras (—1967)
	Assisted by:	
	I. Botelho Conçalves, Horta (Açores)	M. L. da Rocha-Afonso, Lisboa
Romania (Rm)	A. Borza (†), Cluj	N. Boşcaiu, Cluj
Spain (Bl, Hs)	E. F. Galiano, Sevilla	P. Montserrat, Jaca (1971—)
	Assisted by:	
	B. Valdés, Sevilla	
Sweden (Su)	E. Hultén, Stockholm (1967—)	S. Snogerup, Lund (—1967)
Switzerland (He)	O. Hegg, Bern	M. Welten, Bern
Turkey (European part; Tu)	D. A. Webb, Dublin	

U.S.S.R. (Rs (N, B, C, W, K, E))   A. I. Tolmachev, Leningrad
        Assisted by:

A. E. Bobrov, Leningrad	V. P. Klirosova, Kirov
V. I. Czopik, Kiev	L. Laasimer, Tartu
I. J. Fatare, Riga	A. N. Laschenkova, Syktyvkar
T. S. Heidemann, Kishinev	T. Plieva, Leningrad
G. B. Klavina, Riga	I. A. Shabalina, Kirov

CONSULTANTS ON TAXONOMY AND NOMENCLATURE

W. Greuter, Genève	T. Reichstein, Basel
A. C. Jermy, London	J. Sarvela, Helsinki (Dryopteris)
J. D. Lovis, Leeds (Aspleniaceae)	G. Vida, Budapest (Dryopteris)
D. E. Meyer, Berlin	C.-J. Widén, Helsinki (Dryopteris)
R. E. G. Pichi-Sermolli, Genova	

Additional records for the Balkan peninsula have been submitted by P. W. Ball, Clarkson, Ontario, and S. M. Walters, Cambridge.

FINNISH CONSULTATION COMMITTEE, HELSINKI

T. Ahti	I. Kukkonen
P. Isoviita	H. Luther
J. Jalas (Chairman)	R. Ruuhijärvi
A. Kalela	J. Suominen (Secretary)

PREFACE

The body of basic knowledge of vascular plants has increased rapidly in recent decennia, and the taxonomical and geobotanical information scattered throughout the world in innumerable herbaria and botanical papers urgently requires synthesis at international or continental level.

From the point of view of plant geography, including various aspects of chorology, one of the prerequisites of a successful large-scale synthesis of this type is an adequate flora covering the area in question, and providing a common language in respect of the delimitation and nomenclature of the taxa to be treated. This essential condition is now fulfilled for Europe by the current volumes of »Flora Europaea».

Besides standard floras uniting and embodying taxonomical knowledge of local or national scope, carefully prepared and annotated maps showing the distribution of plant species or other taxa in a given area are generally acknowledged to be of the greatest interest and importance for different branches of pure and applied botany. In Europe, especially since the 1950s, plant geographers and others with similar interests have been lucky enough to have access to several high-quality atlases comprising a considerable number of distribution maps for vascular plant species. However, none of these cover the whole of Europe or the total European flora. Distribution maps are also contained by papers scattered throughout the botanical periodicals but the methods of compilation and the degree of accuracy vary, and so far only a small part of the European taxa of vascular plants has been satisfactorily mapped.

While the editors of the atlases mentioned above have achieved admirable results in mapping the distribution of parts of the European flora, the execution of a mapping project intended to cover the whole flora presented in »Flora Europaea» clearly exceeds the capacity of any single person or small national group of botanists. The relatively even coverage of Europe in the maps now presented would certainly not have been possible without the active participation and unflagging interest of all the Committee Members. In fact, it is they, with their expert knowledge of the flora of their respective countries, who should be considered the authors proper of the present Atlas; its whole existence, continuation and scientific standard is entirely dependent on their contributions. The Secretariat is proud to be able to serve as a connecting link within this team. It is the earnest hope of the editors that the Atlas Florae Europaeae will stimulate botanical research in general and chorological studies in particular.

THE EUROPEAN MAPPING SCHEME

It would probably be correct to say that the idea of this Atlas was shaped and inspired by three different achievements. The first was the publication in 1950 of the »Atlas of the Distribution of Vascular Plants in NW. Europe» by Eric Hultén, the Nestor of modern mapping projects, which clearly demonstrated the urgent need of corresponding compilations for areas of similar or even greater size. The second was the publication of the first volume of Flora Europaea in 1964, an indispensable work of reference in matters of nomenclature and taxonomy, but necessarily providing extremely generalized information on distributional patterns.

Thirdly, the rapid development in post-war decades of data-processing methods and similar techniques permitted the development of a modern mapping procedure based on a standard grid system and machine-plotting of the records, and resulted in 1962 in the »Atlas of the British Flora» by F. H. Perring and S. M. Walters.

A map presented by Dr. Perring at the Tenth International Botanical Congress in Edinburgh, 1964, which showed the 50-km square distribution of *Silene acaulis* in Europe, can be regarded as the starting-point of the present work of mapping the flora of Europe. The next step was a mapping experiment, concerning ten species selected by a team nominated at the Edinburgh Congress, and carried out under the direction of Dr. Perring, who, for this purpose, organized a network of collaborators representing various European countries. The results of this experiment were displayed at the Fourth Flora Europaea Symposium in Århus, Denmark (Perring, Bot. Tidsskr. 61: 328—332. 1966).

The Committee for Mapping the Flora of Europe was formally founded at this Symposium, on 10 August 1965. A letter from the Flora Europaea Secretariat had made it clear that, while the closest connexions would be maintained between the two bodies, the Mapping Committee would have to be a separate organization, financed independently. Accordingly, it was agreed that a Secretariat should be set up in Helsinki. Furthermore, a Committee consisting of 14 members and advisers was elected, which included representatives of the major mapping schemes at present in progress in Europe and persons with considerable experience in this work. In December 1965, Dr. J. Suominen took over the secretarial duties of the scheme from Dr. Perring.

At a meeting in Cracow, in September 1966, the Committee discussed the results of the second experimental mapping stage, covering 20 species, together with various details of the mapping techniques, time-table, etc. On the basis of the discussions and decisions of the Cracow meeting, a guide for »Mapping the distribution of European vascular plants» (Memor. Soc. Fauna Fl. Fennica 43: 60—72. 1967) was prepared by the Secretariat and distributed to the members of the team.

At its third meeting, in Halle on 1 July 1968, the Committee examined the progress of the first mapping stage proper, which comprised the species of Pteridophyta and Gymnospermae. It was noted that a number of countries were rather late in sending their records to the Secretariat. This was mainly due to their involvement in local mapping projects, or difficulties experienced by the regional collaborators in establishing a data-collecting network within their respective countries.

It was agreed at the Halle meeting that the membership of the Mapping Committee should be widened to include the whole team, i.e. also the regional collaborators.

Although, for the reasons given above, some data did not reach the Secretariat until December 1971, an eight-page offset proof containing ten species maps (partly incomplete) and the corresponding texts was presented at the Sixth Flora Europaea Symposium in Geneva, in July 1970, and sent to the collaborators for final checking. The entire manuscript on Pteridophyta was finished and distributed by May 1971.

EXPLANATORY NOTES

THE MAPS

Flora Europaea is followed, whenever possible, as to species (and subspecies), and nomenclature. Deviations from this are expressly mentioned. Some slight modifications in the species sequence have been made to facilitate lay-out.

The area mapped is that of Flora Europaea. The Azores and Spitsbergen appear as insets.

The mapping unit is the 50-km square of the UTM (Universal Transverse Mercator) grid maps, covering Europe on a scale of 1 : 1 000 000. Europe comprises ca. 4 400 squares.

The base map, 1 : 10 000 000, shows shorelines, the largest watercourses and political boundaries in black. The positions of the 50-km squares are marked by blue circles with blue code letters, which disappear during printing. This base map [1] is used a) by the Committee Members in the various countries in collecting and sending the 50-km square data and b) by the Secretariat in compiling and publishing the final maps.

The use of a network of collaborators resident in each of the European countries seems to have the following advantages:

Maximum uniformity is obtained in the representation of the various parts of Europe. When obvious gaps in floristic (and taxonomic) knowledge emerge in any part of Europe, the local expert(s) will often make every effort to fill them by organizing the necessary research work.

The most up-to-date material can be gathered by a botanist who is familiar with the local sources of information and has no particular difficulties with the language(s) and place names. Thus even the newest records can be incorporated, including unpublished data from herbaria and elsewhere, and decisions can most easily be made regarding native versus introduced status and age groups of aliens. Certain and possible extinctions can be noted, and erroneous records, sometimes quoted for decades in the literature, can be »killed» effectively. The local experts also participate in the checking of the European maps.

In the base map the 50-km squares are 5 × 5 mm, the mapping symbols measure 4 mm, and the final scale is ca. 1 : 31 000 000.

The 100-km squares of the UTM maps are designated by two letters, and their 50 × 50 km quarters are denoted by the numerals from 1 to 4: $\frac{1}{2}CA\frac{3}{4}$ (see map in the envelop). Along the compensation zones of every sixth meridian, irregular squares occur, their breadth (W-E) varying from 40 to 60 km. The squares are used without any respect to political boundaries. In coastal areas the squares with less than 10 % (250 km²) of land are included in a neighbouring mainland square. However, islands far from the coast and long peninsulas have their own squares. Mapping symbols for isolated islands are put on to the actual position of the island, irrespective of the grid.

---

[1] Printed by »Maanmittaushallituksen karttapaino» (Printing office of the National Board of Survey), Helsinki.

An asterisk (*) indicates symbols of squares the circles of which deviate from the actual position of the UTM squares.

British Isles

The entire S. part of the Isle of Lewis is in PE2.
All of North Uist (etc.) is in *ND3.
All of S. South Uist (Barra, etc.) is in *ND4.
Dunvegan Head is in PD3.
All of the Isle of Man is in *UF4.
All of the Scilly Isles are in PR4.
UA2 is included in *UA1.
UA4 is included in UA3.
Guernsey and Sark are in WV1.
Alderney and Jersey are in WV3.

Denmark

All of Laesö is in *PJ1.
All of Bornholm is in *VB4.
Faeröerne are in *FaeN and *FaeS, the limit between them follows Skopen Fjord.

Germany

Helgoland is in ME1.
VA2 contains only E. Rügen.

Greece

All of Nísos Kerkira is in DJ1.
All of Nísos Thasos is in LF2.
All of Nísos Límnos is in *LE2.
All of Nísos Skiros is in KD2.
E. edge of FK4 is in GK2, SW. corner of FK4 is in FK2.
The higher parts of Psiloriti belong to LU1.

Iceland

All of Snaefellsnes is in *AN4.

Italy

NP4 is included in PP2.
*PN2 contains: all of Elba, Capraia, Pianosa, Montecristo, and islands W. of it. (Mainland of PN2 is included in PN4.)
UF2 contains also islands of UF4.
All of Isole Eolie o Lípari are in *VC3.
All of Pantelleria is in *QF3.
All of Isole Pelagie are in *TV1.
(All of Malta is in *VV1.)

Netherlands

All of Vlieland is in FU1.
All of Terschelling is in FV4.

Norway

Kjelmesöy and the peninsula E. of it are in VC2.
*Björnöya and *Jan Mayen form a square of their own.
Kvitöya is in *NK4.
Kongsöya is in *NH1.
Svensköya is in *MH4.
Hopen is in *MF4.

Portugal

Each island of the *Azores forms a square of its own.

Spain

BD2 is included in *BC1.
All of Isla de Ibiza is in *CD4.
The entire W. part of Isla de Mallorca is in *DD3, the E. part in *ED1.
All of Isla de Menorca is in *EE4.
I. Alboran is in *VE3.

Sweden

All of Gotska Sandön is in *CK3.
All of Central Gotland is in *CJ1.
W. edge of XD1 is in WD3, N. Öland is in XD2.

Turkey (European part)

All of Imroz Ada is in ME2.
PF4 is included in *PF3.

U.S.S.R.

All of Poluostrov Rybachiy is in VC4.
NW. Kanin is in *MB4.
Ostrov Matveyev is in EB3, the rest of EC4 is included in EC3.
All of Sur Sari is in NG1.
Ostrov Malyy Tyutyarsari is in NG2.
Kolka is in EJ3.
Ostrov Dzharylgach is in VS4, Rybnoye in VR4.

Any reliable record was used when entering a symbol in a square, not only those based on herbarium specimens but also reliable published or unpublished sight records. Frequency or abundance within the square are not indicated. However, some other particulars are given.

The mapping symbols used are:

● native occurrence

○ introduction (established alien)

◐ status unknown or uncertain

(Omitted: rare and/or ephemeral casuals; short-distance and/or inconstant escapes from cultivation; cultivations, even large-scale cultivations of forest trees.)

✚ extinct

✖ probably extinct

? record uncertain (as to identification or locality)

The age of records for introduced species, especially those showing expanding tendencies, are sometimes indicated:

◓ before 1900

◒ 1900—1939

◑ 1940 onwards

Within the same square, native occurrences have been given preference over introductions, extant over extinct, certain over uncertain, old records of expanding species over more recent ones.

The way of using the different special symbols varies somewhat from country to country. Thus in Br, the symbol ✖ (probably extinct) means any record made before 1930.

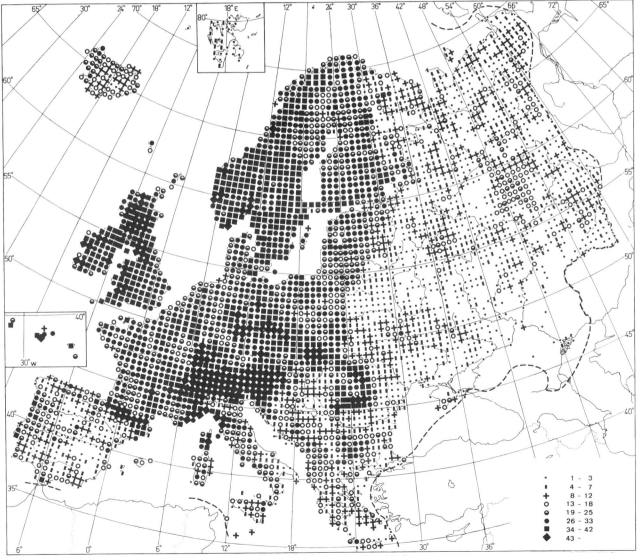

The numbers of species and subspecies of Pteridophytes in the 50-km squares.

For the second and third taxon in a map the symbols ▲ and ✳ are used. The special symbols (○, ◐, ✚, ✖, ?, etc.) belong to the taxon marked by ●, if not otherwise stated.

The map (p. 11) giving the number of species of *Pteridophyta* recorded for each 50-km square shows the great differences in the richness of the flora within Europe, and draws attention to the squares with the greatest variety of habitats and vegetation zones. It also indicates the differences in the relative intensity of floristic research.

Geological, climatic and biogeographical information on Europe can be found in various handbooks, and is excluded here. However, information on the maximum and minimum altitudes in each 50-km square is given in the maps below. These maps may help in the interpretation of the records of lowland and mountain species.

THE TEXTUAL COMMENTS

The names of the species and the author(s) are followed by a reference to the number in a corner of the map concerned. Important synonyms, especially those appearing in the text of Flora Europaea, are given in 8-point type.

The subsequent headings in *italics* begin paragraphs giving special notes on taxonomy, nomenclature, and important new or omitted records. As a rule, these are given only when they complete or correct

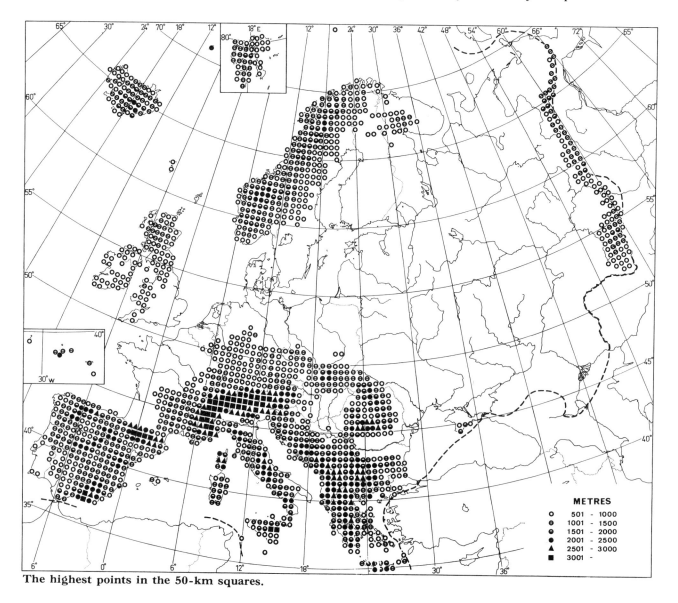

METRES

○	501 - 1000
◓	1001 - 1500
◑	1501 - 2000
●	2001 - 2500
▲	2501 - 3000
■	3001 -

The highest points in the 50-km squares.

Flora Europaea, or are necessary in explaining some special difficulty in mapping and its solution. A collective map was often the only possibility for difficult species (groups), even if the taxa in question are given separately in Flora Europaea and have separate lists of the countries in which they occur, and even if they could be mapped separately in some of the countries.

The information under *total range* mainly comprises references to maps where the reader can study this question, which is often fairly complicated and difficult to explain shortly and exactly. Problems exist even in cases where the range is well studied, particularly owing to difficulties with the delimitation of the species.

Abbreviations are used for the most frequently quoted literature:

Fl. Eur. = T. G. Tutin, V. H. Heywood et al. (ed.), Flora Europaea 1. — xxxiv + 464 pp. Cambridge 1964.

Hultén AA 1958 = E. Hultén, The Amphi-Atlantic Plants. — Kungl. Svenska Vet.-Akad. Handl., Ser. 4, 7(1): 1—340. 1958.

Hultén CP 1964 = E. Hultén, The Circumpolar Plants. I. — Kungl. Svenska Vet.-Akad. Handl., Ser. 4, 8(5): 1—280. 1964.

MJW 1965 = H. Meusel, E. Jäger & E. Weinert, Vergleichende Chorologie der zentraleuropäischen Flora. — Text 583 pp., Karten 258 pp. Jena 1965.

Other literature is quoted by giving its author(s) and the periodical in which it appeared, with volume, page(s), and year of publication.

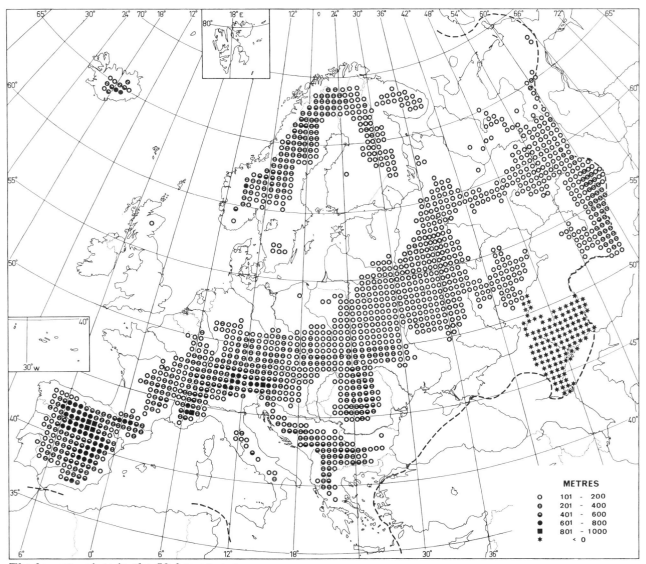

METRES

○	101 - 200
◉	201 - 400
◕	401 - 600
●	601 - 800
■	801 - 1000
✳	< 0

The lowest points in the 50-km squares.

ACKNOWLEDGEMENTS

The fruitful contacts maintained with the Flora Europaea organization has been of the utmost value to the Committee for Mapping the Flora of Europe, especially in the early stages of its work. All the mapping material accumulated before the foundation in 1965 of the present mapping committee was generously placed at our disposal by the Flora Europaea Secretariat and Dr. F. H. Perring of the Biological Records Centre, Monks Wood Experimental Station, England. The progress reports on the mapping of the flora of Europe regularly given at the Flora Europaea Symposia bear witness to the close connexions between these parallel undertakings.

Right from the beginning of the activities of the Secretariat, the Finnish Academy of Science and Letters showed active interest in the mapping project. Among other things, a commission was appointed to present the necessary financial motions to the governmental organs. The Secretariat's work in the first year of its activity, 1966, was financed by a grant from the Finnish National Research Council for Sciences, and from 1967 onwards its fairly modest annual expediture has been covered by means put at its disposal by the Ministry of Education of Finland.

We highly appreciate the readiness of the Department of Botany, University of Helsinki (Head: Professor Aarno Kalela), to provide working facilities for the Secretariat, although troubled by a chronic lack of space.

As mentioned before, the expert work done by the individual Committee Members and other participants in gathering the material for their respective countries is a fundamental condition for the success of a project of the scale of the European mapping scheme. Our respectful thanks are also due to the Learned Academies and Societies, University Departments and Herbaria throughout Europe that have forwarded the work of the Committee Members and their local collaborators in various ways. We are particularly grateful to the Polish Academy of Sciences and the University of Cracow, as well as to the Biological Society of the German Democratic Republic and the Martin-Luther University of Halle-Wittenberg, for their kindness in inviting the Committee to meet in Cracow and Halle, respectively.

At times, the main burden of work in the Secretariat has fallen on the technical assistant, Mrs. Ulla Kurimo, M.A., whose skill and care in preparing the final maps is highly appreciated. The Secretariat is also much obliged to Mrs. Aune Koponen, M.A., who prepared the base map, as well as to the staff of the Department of Botany, University of Helsinki, and all the other persons who participated in the work.

Last but not least, our sincere gratitude is due to the Finnish Biological Society Vanamo, who lifted a great weight off our minds, when they informed us of their decision to undertake the publishing of Atlas Florae Europaeae.

PTERIDOPHYTA

PSILOPSIDA

PSILOTACEAE

Psilotum nudum (L.) Beauv. — Map 1.

Psilotum triquetrum (L.) Swartz

Found in 1965, first record from Europe (B. Molesworth Allen, Taxon 15: 82—83 (1966), Brit. Fern Gaz. 9: 249—250 (1966); see also F. Bellot Rodriguez, Anal. R. Acad. Farm. (Madrid) 32: 417 —423 (1966)).

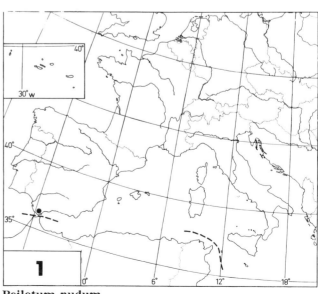

Psilotum nudum

LYCOPSIDA

LYCOPODIACEAE

Huperzia selago (L.) Bernh. ex Schrank & Mart.

Lycopodium selago L.

Notes. Co added (not given in Fl. Eur.), Rs(E) omitted (given in Fl. Eur.), Sb Björnöya omitted (given in Hultén CP 1964).

Total range. Hultén CP 1964: map 46; MJW 1965: map 8a.

H. selago subsp. **selago** — Map 2.

In the north the records partly refer to subsp. *arctica;* see below.

H. selago subsp. **arctica** (Grossh.) Á. & D. Löve.

Not mapped separately. In addition to Spitsbergen (Fl. Eur.), recorded from northern Russia, cf. A. I. Tolmachev, Arkt. Fl. SSSR. 1: 55 (1960). Its status and its relationship to *H. selago* subsp. *appressa* (Desv.) D. Löve are unclear.

H. selago subsp. **dentata** (Herter) Valentine — Map 2.

Huperzia selago subsp. suberecta Franco & Vasc., Bol. Soc. Brot. 41: 23 (1967), non Lycopodium suberectum Lowe

Nomenclature. The type specimen of *Lycopodium suberectum* Lowe (in BM) belongs to *H. selago* subsp. *selago* (G. Benl, Brit. Fern Gaz. 10: 172 (1971)).

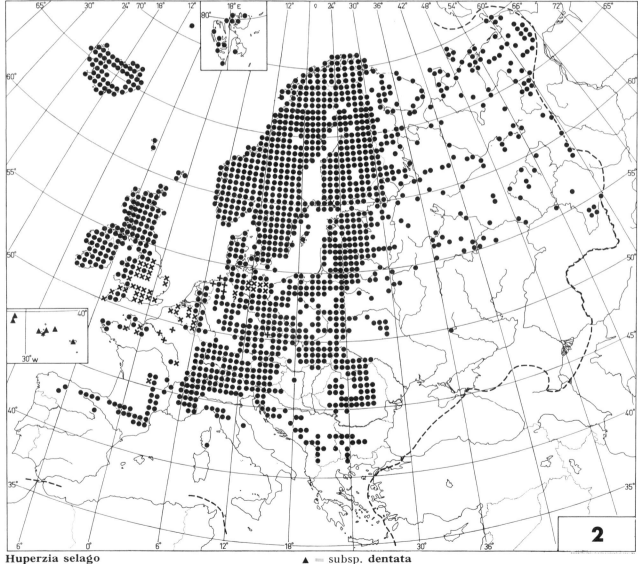

Huperzia selago
● = subsp. **selago**

▲ = subsp. **dentata**
✱ = subsp. **selago** & subsp. **dentata**

Lycopodiella cernua (L.) Pichi-Serm. — Map 3.

Lepidotis cernua (L.) Beauv.; Lycopodium cernuum L.; Palhinhaea cernua (L.) Franco & Vasc.

Nomenclature and generic delimitation. J. Vasconcellos & J. do Amaral Franco, Bol. Soc. Brot. 41: 24—25 (1967); R. E. G. Pichi-Sermolli, Webbia 23: 166 (1968).

Total range. Pantropical. H. Meusel & J. Hemmerling, Die Bärlappe Europas, p. 26 (Wittenberg Lutherstadt 1969).

Lycopodiella inundata (L.) J. Holub — Map 4.

Lepidotis inundata (L.) C. Börner; Lycopodium inundatum L.

Nomenclature and generic delimitation. J. Holub, Preslia 36: 16—22 (1964).

Notes. Az added (not given in Fl. Eur.), Hu omitted (given in Fl. Eur.), Rs(E) added (not given in Fl. Eur.).

Total range. Hultén AA 1958: map 198; MJW 1965: map 8b.

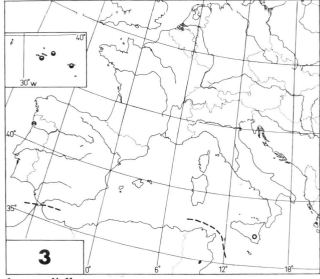

Lycopodiella cernua

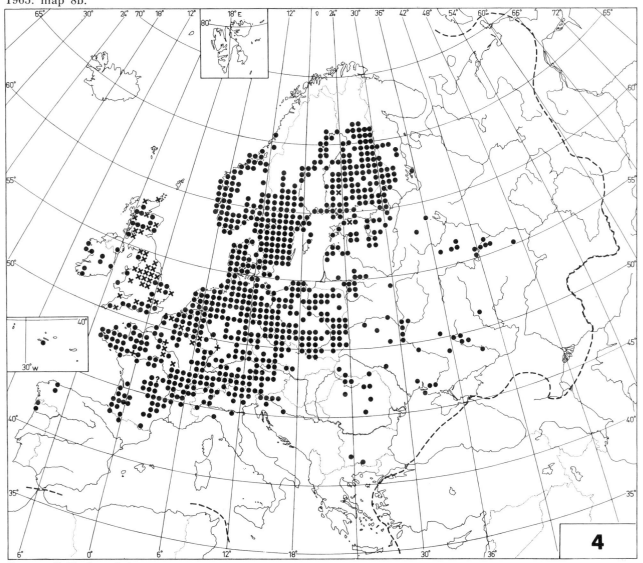

Lycopodiella inundata

Lycopodium annotinum L. — Map 5.

Incl. Lycopodium dubium Zoega & L. pungens (La Pylaie) Iljin

Taxonomy. Intermediates between *L. dubium* and *L. annotinum* are numerous and, accordingly, the former may only deserve subspecific or varietal rank under *L. annotinum;* cf. P. Kallio et al., Ann. Univ. Turkuensis A II 42 (Rep. Kevo Subarctic Res. Stat. 5): 56—58 (1969). Subspecific names to be considered are *L. annotinum* subsp. *alpestre* (Hartm.) Á. & D. Löve, Nucleus 1: 7 (1958), and *L. annotinum* subsp. *pungens* (La Pylaie) Hultén, Ark. Bot. 7: 7 (1968).

Notes. Bu and Hs omitted (given in Fl. Eur.).

Total range. Hultén CP 1964: map 62.

Lycopodium annotinum
Line = approximate S. limit of **L. dubium**

Lycopodium clavatum L. — Map 6.

Taxonomy. Var. *lagopus* Laest. (subsp. *monostachyon* (Grev. & Hook.) Selander) perhaps deserves subspecific rank but has not been mapped separately.

Total range. Hultén CP 1964: map 61.

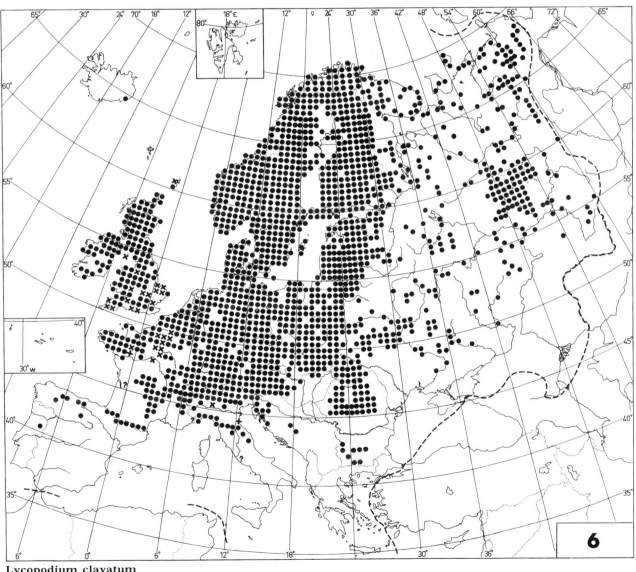

Lycopodium clavatum

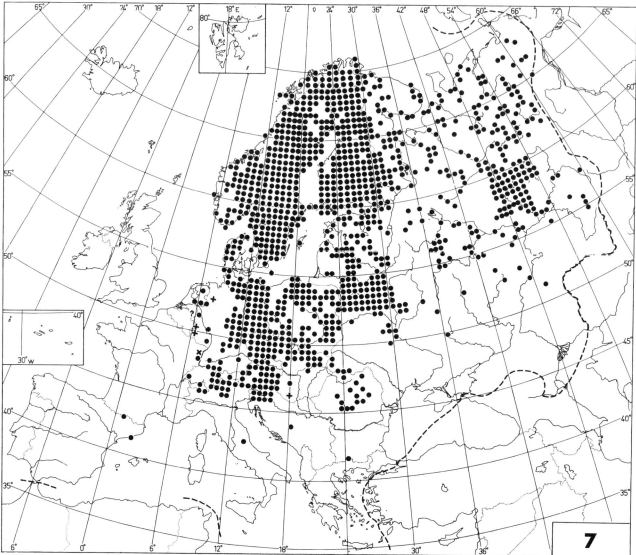

Diphasium complanatum subsp. **complanatum**

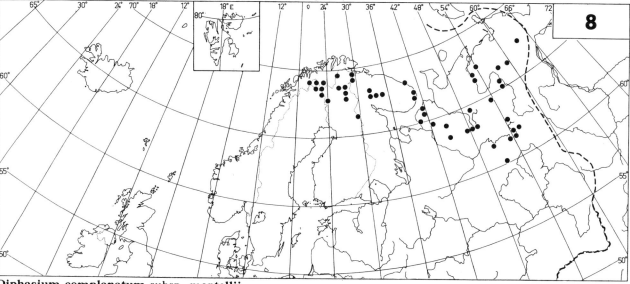

Diphasium complanatum subsp. **montellii**

Diphasium complanatum (L.) Rothm.

Lycopodium complanatum L.; incl. Diphasium tristachyum (Pursh) Rothm.

Taxonomy. I. Kukkonen, Ann. Bot. Fenn. 4: 441—470 (1967), 7: 142 (1970).

Total range. MJW 1965: map 8c.

D. complanatum subsp. complanatum — Map 7.

Lycopodium anceps Wallr.

Notes. In Be extinct (given as present in Fl. Eur.).

Total range. Hultén CP 1964: map 109; J. Wilce, Beih. Nova Hedwigia 19: pl. 34 (1965), somewhat incomplete. Both include subsp. *montellii*.

D. complanatum subsp. montellii Kukk. — Map 8.

Taxonomy and nomenclature. I. Kukkonen, Ann. Bot. Fenn. 4: 441—470 (1967), 7: 142 (1970). Intermediates between subsp. *montellii* and subsp. *complanatum* occur.

Notes. The inclusion in this subspecies of the material from northern Russia is in accordance with recent studies by I. Kukkonen (unpubl.).

D. complanatum subsp. chamaecyparissus (A. Braun) Kukk. — Map 9.

Diphasium tristachyum (Pursh) Rothm.; Lycopodium chamaecypasissus A. Braun; L. complanatum subsp. chamaecyparissus (A. Braun) Nyman

Taxonomy. Intermediates between subsp. *chamaecyparissus* and subsp. *complanatum* (*D.* × *zeilleri* (Rouy) Damboldt) are present in at least Au, Be, Cz, Da, Fe, Ga, Ge, Ho, No, Po, Rs(B, C) and Su; cf. S. Rauschert, Hercynia 4: 462—471 (1967).

Notes. Occurrence in Au uncertain (given as certain in Fl. Eur.). Records in Hultén AA 1958, etc., from northern Russia refer to subsp. *montellii*.

Total range. Hultén AA 1958: map 39; J. Wilce, Beih. Nova Hedwigia 19: 144, pl. 35 (1965).

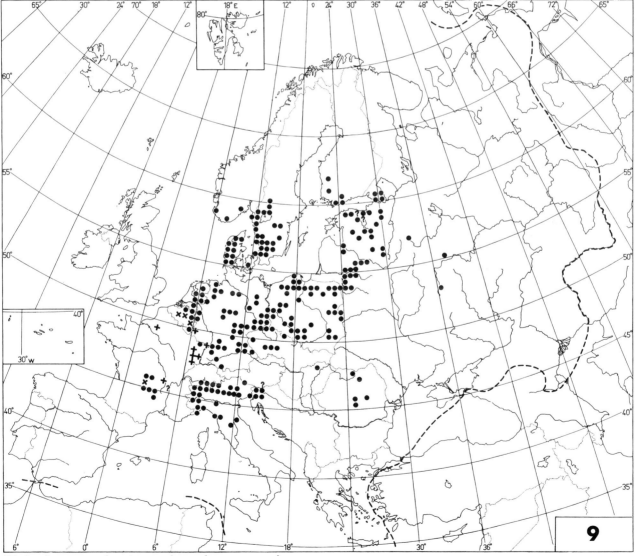

Diphasium complanatum subsp. **chamaecyparissus**

22

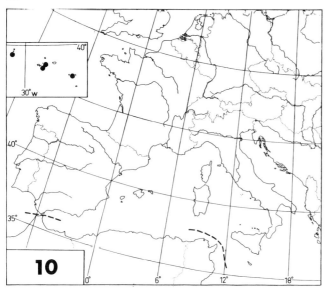

Diphasium madeirense (Wilce) Rothm. —Map 10.

Lycopodium madeirense Wilce

Nomenclature. W. Rothmaler, Feddes Repert. 66: 235 (1962).

Total range. Açores and Madeira; J. Wilce, Beih. Nova Hedwigia 19: 140, pl. 38 (1965). Records included with those of *D. complanatum* in Hultén CP 1964: map 109, and MJW 1965: map 8c.

Diphasium madeirense

Diphasium issleri (Rouy) J. Holub — Map 11.

Lycopodium issleri (Rouy) Lawalrée

Taxonomy. According to J. Wilce, Beih. Nova Hedwigia 19: 93 (1965), a hybrid derivative of *D. alpinum* and *D. complanatum* subsp. *chamaecyparissus*.

Notes. Br and Hu omitted (given in Fl. Eur.), He added (not given in Fl. Eur.).

Total range. Endemic to Europe (Fl. Eur.). However, J. Wilce, Beih. Nova Hedwigia 19: 93, pl. 33 (1965), gives one station in North America.

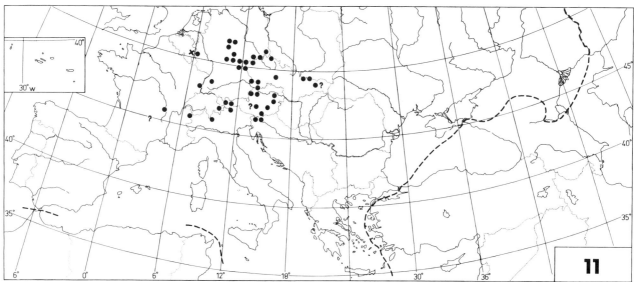

Diphasium issleri

Diphasium alpinum (L.) Rothm.   Map 12.

Lycopodium alpinum L.

Notes. In Be extinct (given as present in Fl. Eur.), Da confirmed (?Da in Fl. Eur.), Rs(W) added (not given in Fl. Eur.).

Total range. Hultén AA 1958: map 215; MJW 1965: map 8d.

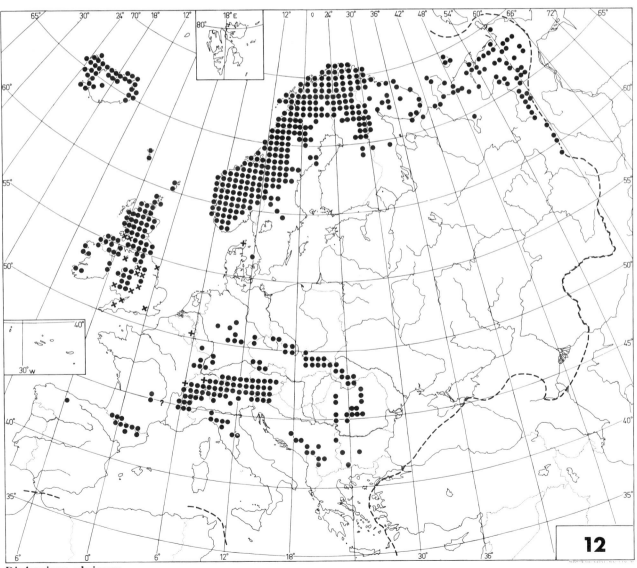

Diphasium alpinum

SELAGINELLACEAE

Selaginella selaginoides (L.) Link — Map 13.

Selaginella spinulosa A. Braun

Notes. Hb added (not given in Fl. Eur.).
Total range. Hultén AA 1958: map 222; MJW 1965: map 9a.

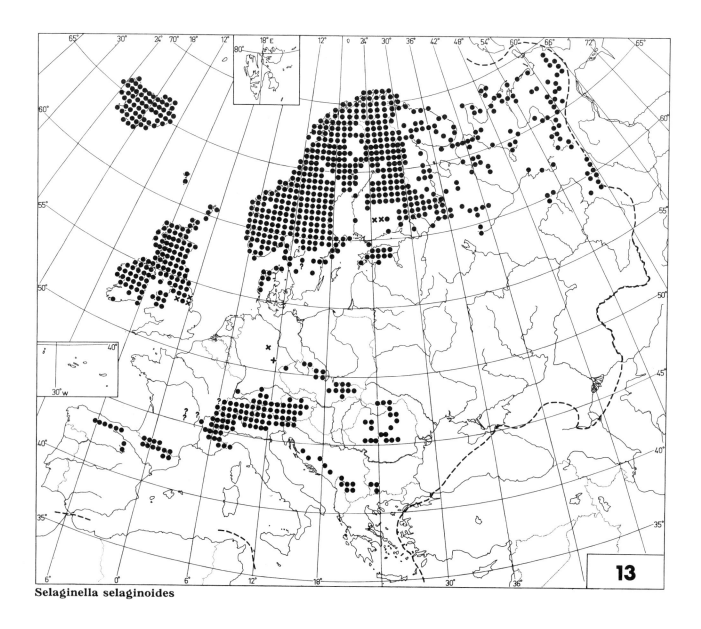

Selaginella selaginoides

13

Selaginella helvetica (L.) Spring — Map 14.

Notes. In Be probably not indigenous and now extinct (given as native and present in Fl. Eur.), Rs(W) added (not given in Fl. Eur.). The dubious record of *S. denticulata* from Tu perhaps belongs here (D. A. Webb, Proc. Roy. Irish Acad. 65 B 1: 9 (1966).

Total range. MJW 1965: map 9b.

Selaginella denticulata (L.) Link — Map 15.

Notes. For Tu, cf. under *S. helvetica*.

Total range. R.E.G. Pichi-Sermolli, Lav. Soc. Ital. Biogeogr. 1: 107 (1971).

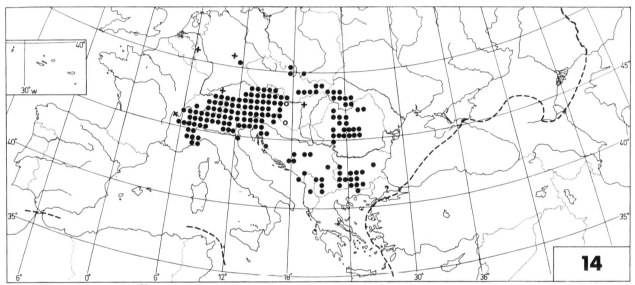

Selaginella helvetica

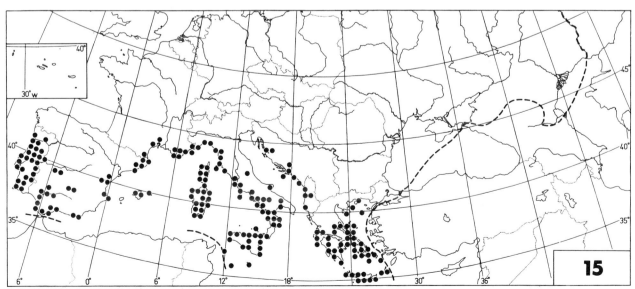

Selaginella denticulata

Selaginella apoda (L.) Spring — Map 16.

Nomenclature. C. V. Morton, Amer. Fern Jour. 57: 106 (1967).

Notes. Since 1869 known as naturalized in one locality in Berlin; see D. E. Meyer, Berliner Naturschutzbl. 14 (Sonderh.): 16—18 (1970).

Native of North America.

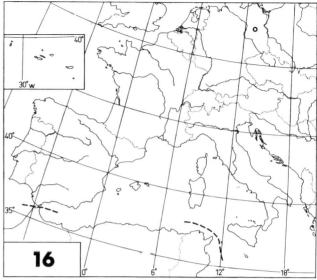

16

Selaginella apoda

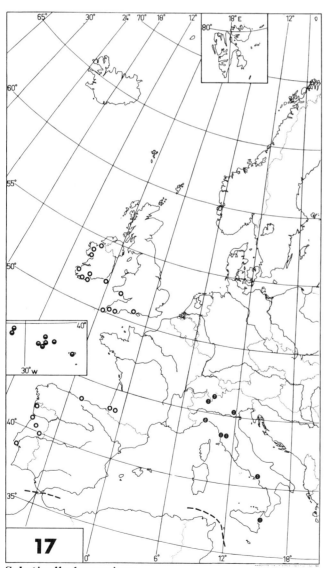

17

Selaginella kraussiana

Selaginella kraussiana (G. Kunze) A. Braun — Map 17.

Selaginella azorica Baker

Notes. In Be extinct, Ge and Hs omitted ([Be, Ge, Hs] in Fl. Eur.).

Native of tropical and south Africa.

ISOËTACEAE

Isoëtes lacustris L. — Map 18.

Notes. Be and Bu omitted (given in Fl. Eur.), Fa added (not given in Fl. Eur.), Gr omitted (?Gr in Fl. Eur.), It omitted (given in the text of Fl. Eur.). The only record from Rm is old and unconfirmed, and the habitat has been destroyed.

Total range. Hultén AA 1958: map 247; MJW 1965: map 9c. If taken in the narrow sense, endemic to Europe (as given in Fl. Eur.), with the exception of one station in south Greenland.

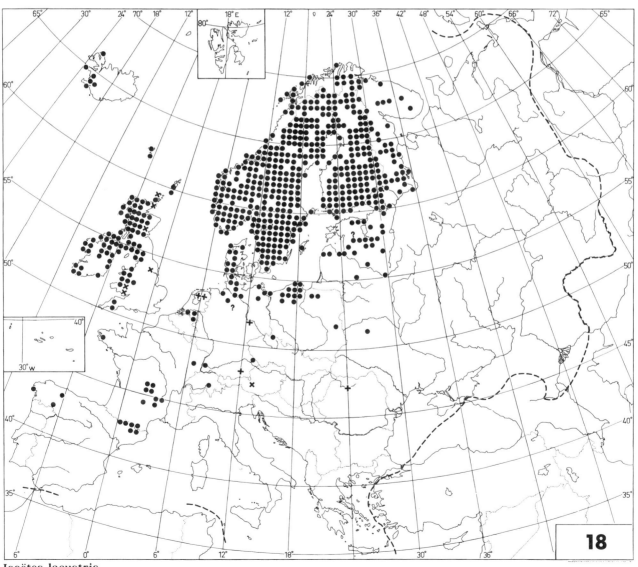

Isoëtes lacustris

28

Isoëtes echinospora Durieu — Map 19.

Isoëtes tenella Léman ex Desv.; I. setacea auct., non Lam.

Taxonomy. The Icelandic plants are often referred to the closely allied North American *I. braunii* Durieu (*I. echinospora* subsp. *muricata* (Durieu) Á. & D. Löve var. *braunii* (Durieu) Á. & D. Löve).

Nomenclature. A. C. Jermy, Brit. Fern Gaz. 10: 106 (1969).

Notes. Au omitted (given in Fl. Eur.), Be, Bu, Fa and Gr added (not given in Fl. Eur.), Hs confirmed (?Hs in Fl. Eur.), Rm omitted (?Rm in Fl. Eur.).

Total range. Hultén AA 1958: map 235; MJW 1965: map 9d.

Isoëtes brochonii Moteley — Map 20.

Notes. Hs confirmed (?Hs in Fl. Eur.). Occurrence in Ga uncertain (given as certain in Fl. Eur.), the records perhaps referring to juvenile *I. echinospora*.

Total range. Endemic to Europe.

Isoëtes azorica Durieu ex Milde — Map 21.

Total range. Endemic to the Açores.

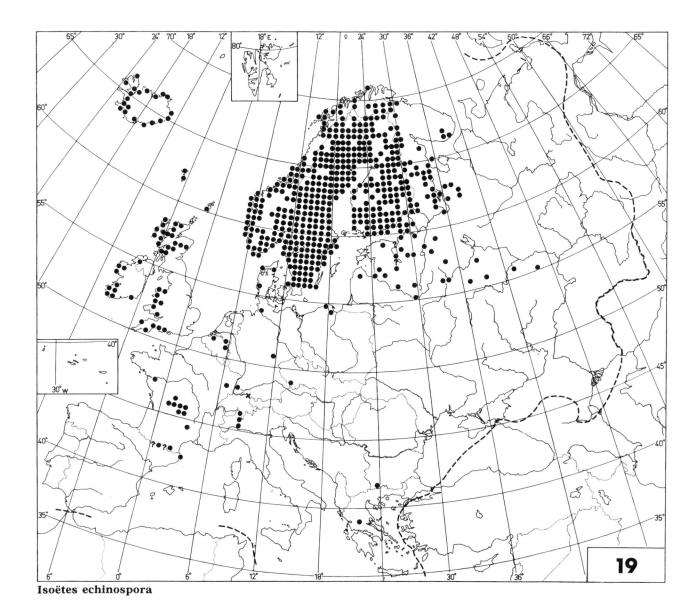

Isoëtes echinospora

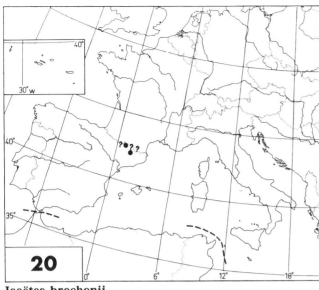

Isoëtes brochonii

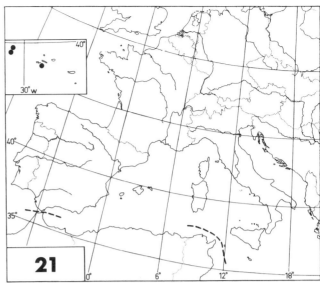

Isoëtes azorica

Isoëtes boryana Durieu

Notes. Hs confirmed (?Hs in Fl. Eur.).
Total range. Endemic to Europe.

I. boryana subsp. **boryana** —- Map 22.

I. boryana subsp. **asturicensis** Laínz — Map 22.

Taxonomy. M. Laínz, Bol. Inst. Estud. Astur. (Supl. Ci.) 15: 6—7 (1970).

Isoëtes delilei Rothm. — Map 23.

Isoëtes setacea Bosc ex Delile, non Lam.

Notes. ?Co added (not given in Fl. Eur.).
Total range. Endemic to Europe.

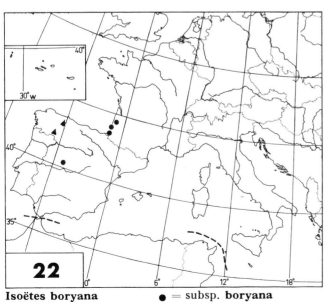

Isoëtes boryana   ● = subsp. **boryana**
       ▲ = subsp. **asturicensis**

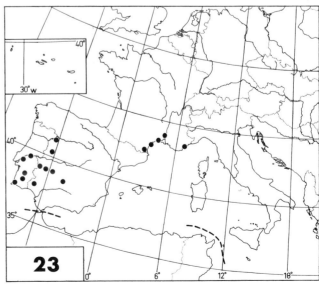

Isoëtes delilei

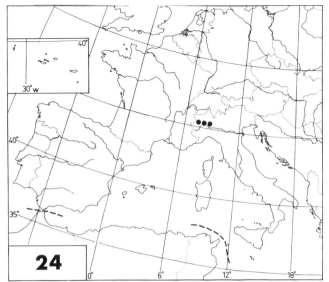

24

Isoëtes malinverniana

Isoëtes malinverniana Cesati & De Not. — Map 24.

Total range. Endemic to Europe.

Isoëtes heldreichii Wettst. — Map 25.

   Total range. Endemic to Europe.

Isoëtes velata A. Braun — Map 26.

   Isoëtes baetica Willk.; I. variabilis Le Grand

Isoëtes tegulana Genn. — Map 26.

   Isoëtes velata var. tegulensis (Genn.) Fiori

   Total range. Endemic to Sardegna.

Isoëtes tenuissima Boreau — Map 27.

   Isoëtes viollaei Hy

   Total range. Endemic to Europe.

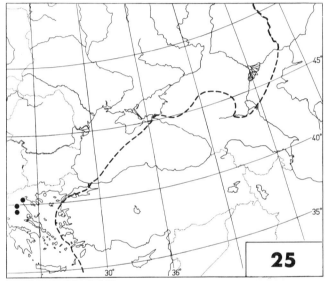

25

Isoëtes heldreichii

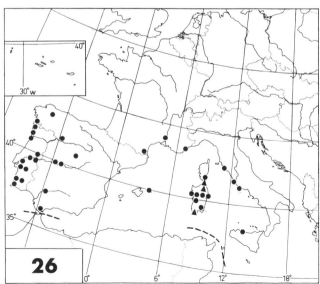

26

● = **Isoëtes velata**
▲ = **I. velata & I. tegulana**

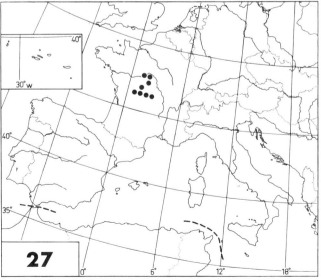

27

Isoëtes tenuissima

Isoëtes histrix Bory — Map 28.

Isoëtes delalandei Lloyd; I. phrygia Hausskn.

Isoëtes durieui Bory — Map 29.

Notes. Gr added (not given in Fl. Eur.).

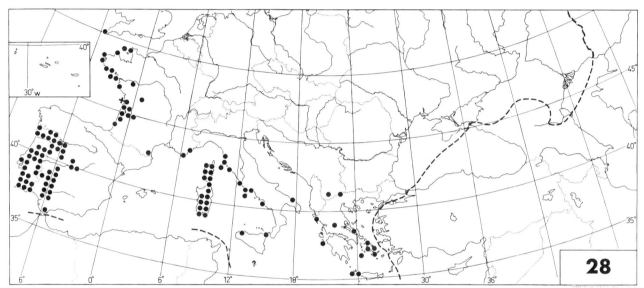

Isoëtes histrix

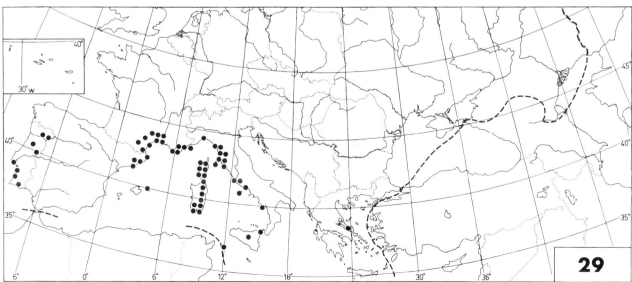

Isoëtes durieui

SPHENOPSIDA

EQUISETACEAE

Equisetum hyemale L. — Map 30.

Notes. Fa added (not given in Fl. Eur.), Cr and Lu omitted (given in Fl. Eur.), the records referring to *E. ramosissimum*, Rs(K) omitted (given in Fl. Eur.).

Total range. Hultén CP 1964: map 174.

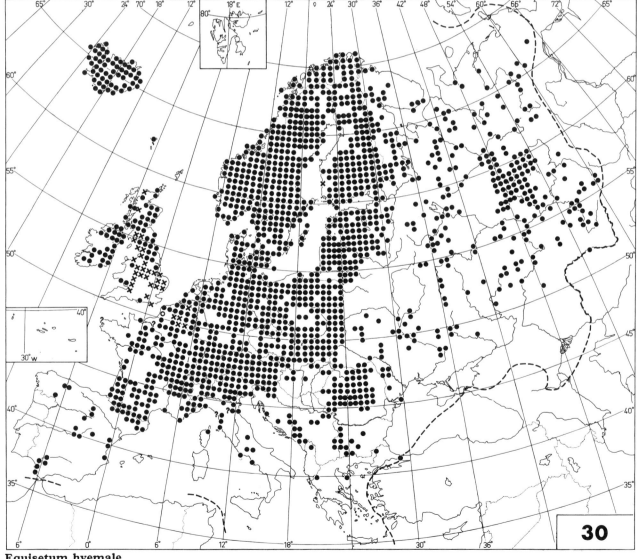

Equisetum hyemale

Equisetum ramosissimum Desf. — Map 31

Notes. Be omitted (given in Fl. Eur.), perhaps not native in Br, probably extinxt or not native in Rs(B) (given as native in Fl. Eur.).

Total range. MJW 1965: map 7d; R. E. G. Pichi-Sermolli, Lav. Soc. Ital. Biogeogr. 1: 96 (1971).

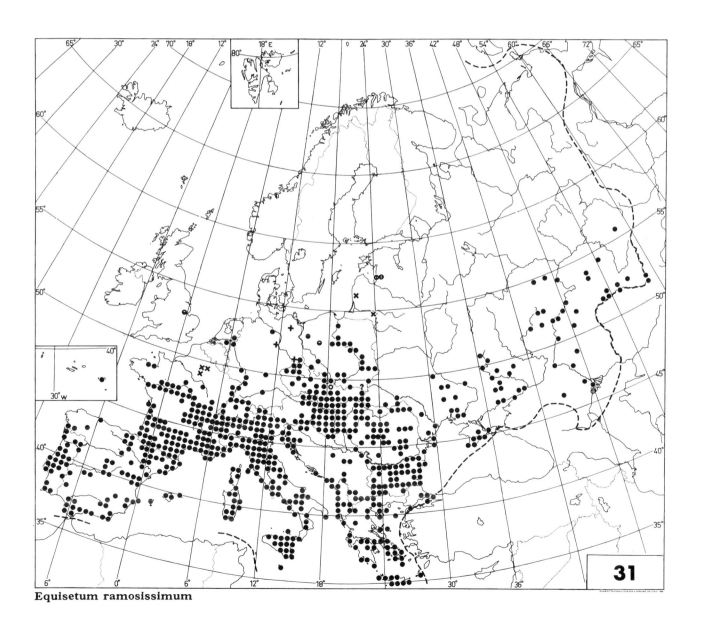

Equisetum ramosissimum

Equisetum variegatum Schleicher ex Weber & Mohr — Map 32.

Notes. Fa and Rs(W) added (not given in Fl. Eur.), Co omitted (given in Fl. Eur.). A number of low-altitude records from Ga are in need of confirmation.

Total range. Hultén CP 1964: map 45.

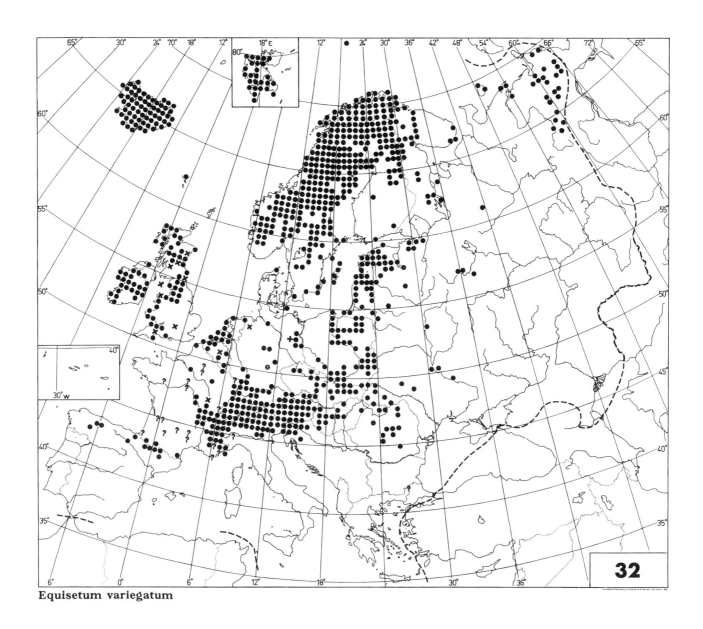

Equisetum variegatum

Equisetum scirpoides Michx — Map 33.

Notes. Au omitted (?Au in Fl. Eur.), Is omitted (given in Hultén CP 1964), Rs(E) added (not given in Fl. Eur.).
Total range. Hultén CP 1964: map 30.

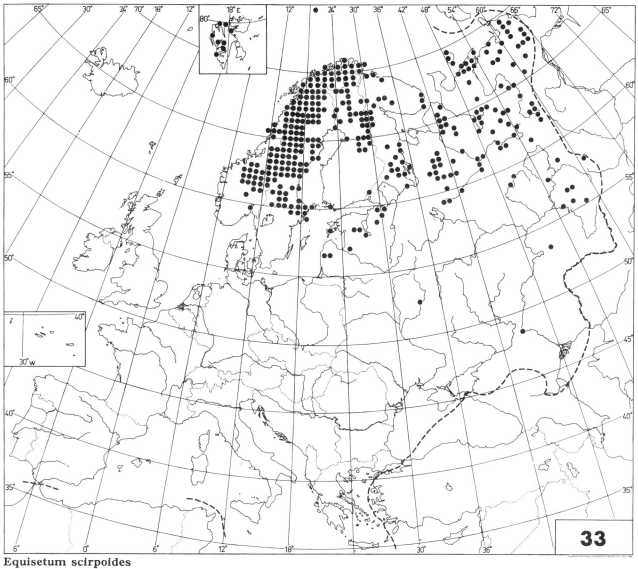

Equisetum scirpoides

Equisetum fluviatile L. — Map 34.

Notes. Bl omitted (given in Hultén CP 1964), Gr added (not given in Fl. Eur.), Rs(K) omitted (given in Fl. Eur.).
Total range. Hultén CP 1964: map 96.

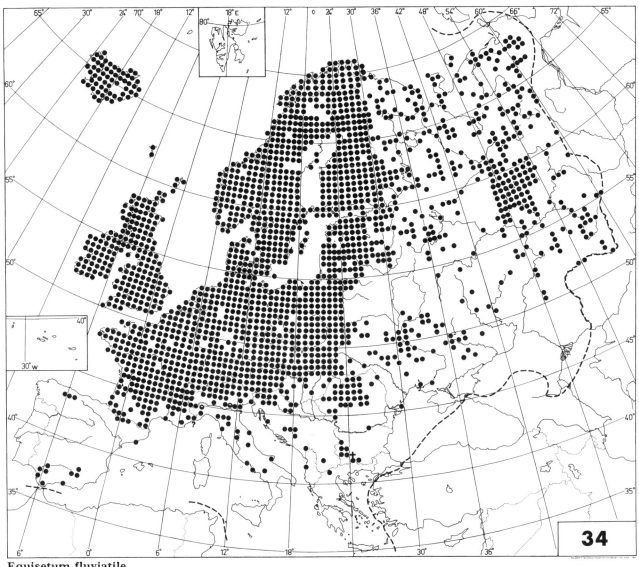

Equisetum fluviatile

Equisetum palustre L. — Map 35.

Notes. In Cr and Si doubtful (given as certain in Fl. Eur.), the records probably referring to *E. ramosissimum*, Rs(K) added (not given in Fl. Eur.), Sa omitted (given in Fl. Eur.).

Total range. Hultén CP 1964: map 89.

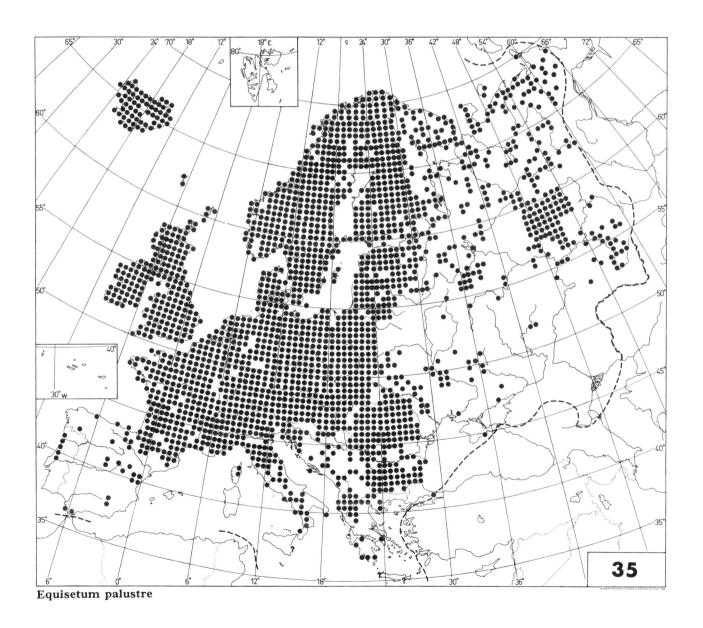

Equisetum palustre

38

Equisetum sylvaticum L. — Map 36.

Notes. Cr omitted (given in Fl. Eur.), Cz confirmed (?Cz in Fl. Eur.), Rs(E) added (not given in Fl. Eur.), Tu omitted (?Tu in Fl. Eur.).

Total range. Hultén CP 1964: map 86; MJW 1965: map 7a.

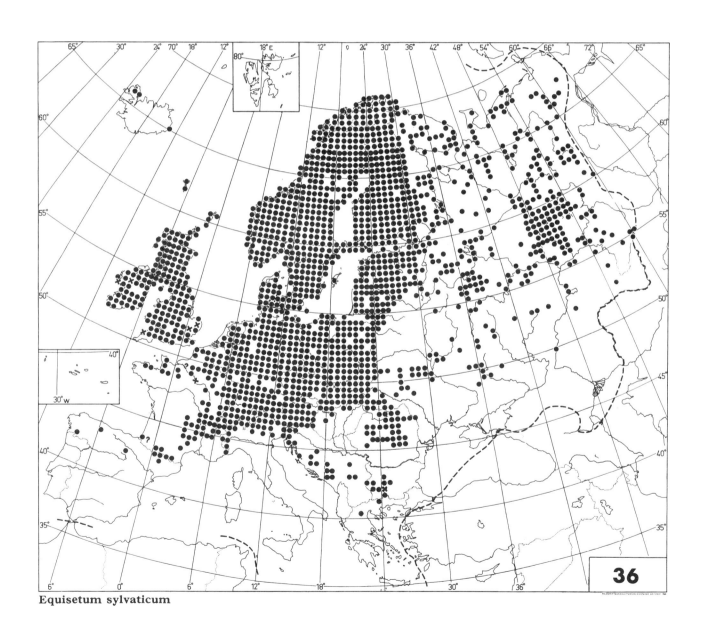

Equisetum sylvaticum

Equisetum pratense Ehrh. — Map 37.

Notes. In Be and Ga uncertain (given as certain in Fl. Eur.), Fa, Ju and Rm added (not given in Fl. Eur.), Rs(K) omitted (given in Fl. Eur.).

Total range. Hultén CP 1964: map 83.

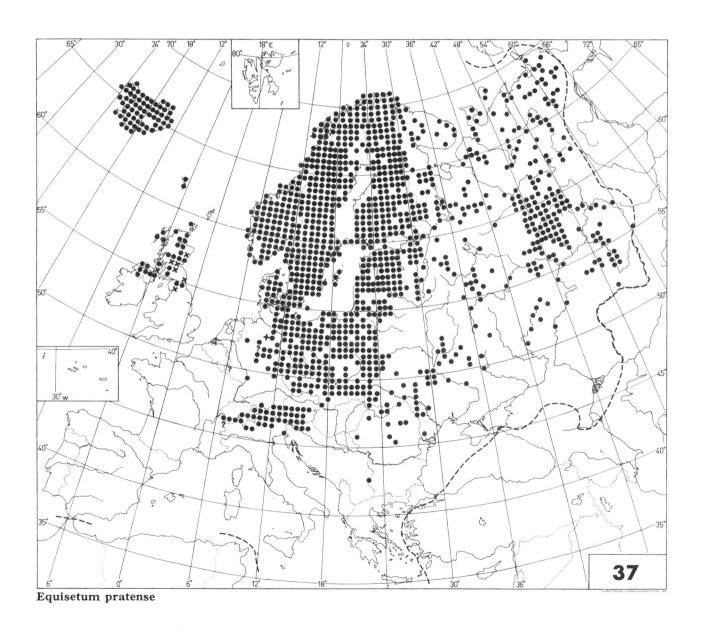

Equisetum pratense

40

Equisetum arvense L. — Map 38.

Total range. Hultén CP 1964: map 98; MJW 1965: map 7c.

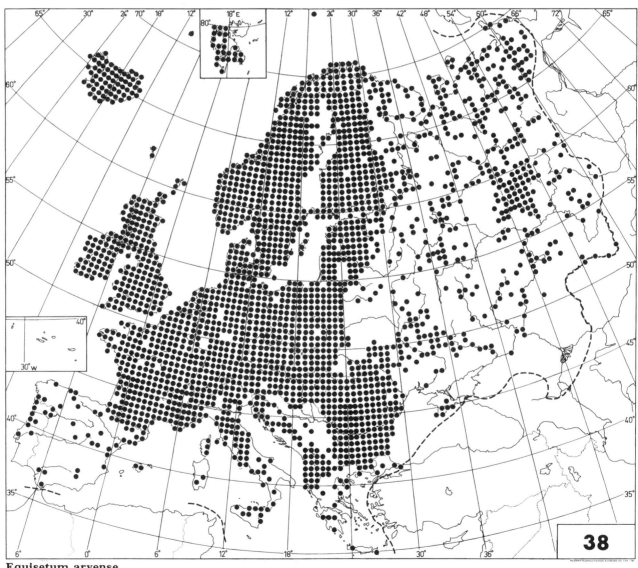

Equisetum arvense

Equisetum telmateia Ehrh. — Map 39.

Equisetum majus Gars.; E. maximum auct.

Total range. Hultén AA 1958: map 258; MJW 1965: map 7b; E. Jäger, Feddes Repert. 79: 228 (1968).

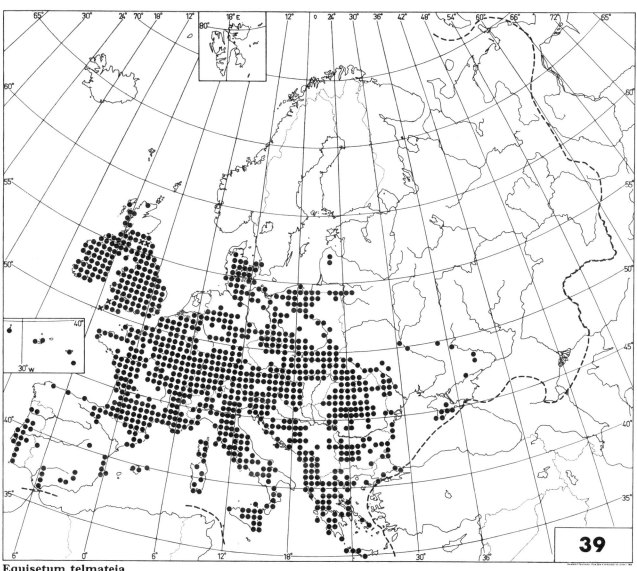

Equisetum telmateia

FILICOPSIDA

OPHIOGLOSSACEAE

Ophioglossum lusitanicum L. — Map 40.

> *Notes.* Az confirmed (?Az in Fl. Eur.).
> *Total range.* R. T. Clausen, Mem. Torrey Club 19 (2): 145 (1938).

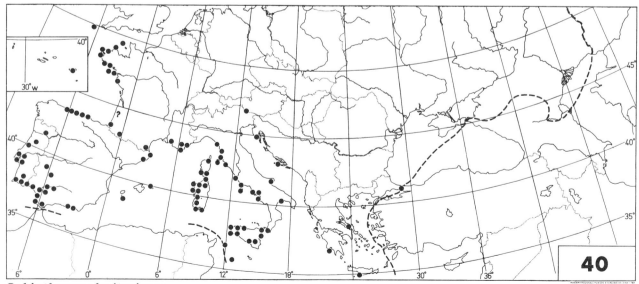

Ophioglossum lusitanicum

Ophioglossum azoricum C. Presl — Map 41.

Ophioglossum vulgatum L. subsp. ambiguum (Cosson & Germ.) E. F. Warburg; O. vulgatum L. subsp. polyphyllum auct., non var. polyphyllum A. Braun

Taxonomy. An alloploid (2n = 720), derived from *O. lusitanicum* and *O. vulgatum;* Á. Löve & B. M. Kapoor, Nucleus 9: 132—138 (1966).

Notes. Co added (not given in Fl. Eur.), Hs omitted (given in Fl. Eur.).

Total range. Endemic to Europe (Fl. Eur.). However, G. Kunkel, Cuad. Bot. (Gran Canaria) 3: 55—56 (1968) reports it as native for Islas Canarias.

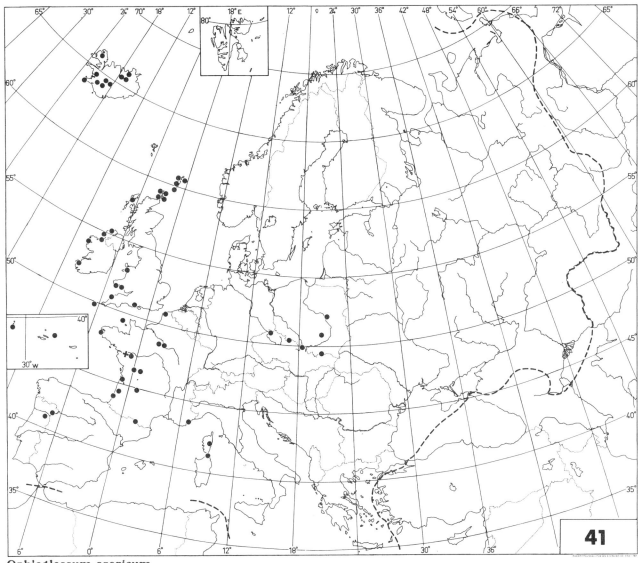

41

Ophioglossum azoricum

44

Ophioglossum vulgatum L. — Map 42.

Notes. Cr omitted (given in Fl. Eur.).

Total range. Hultén CP 1964: map 91; MJW 1965: map 10a. The maps are collective, including *O. azoricum*, at least.

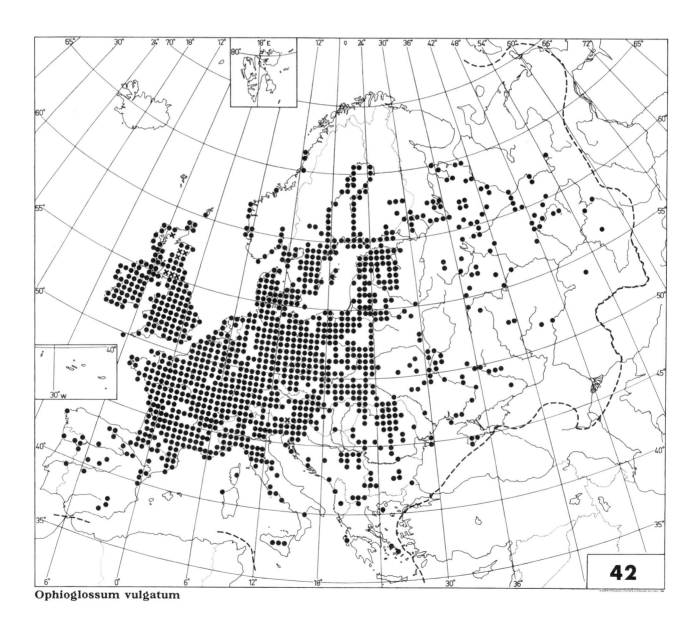

Ophioglossum vulgatum

Botrychium simplex E. Hitchc. — Map 43.

Notes. Be omitted (given in Fl. Eur.), Ge added (not given in Fl. Eur.).
Total range. Hultén AA 1958: map 193; MJW 1965: map 11a.

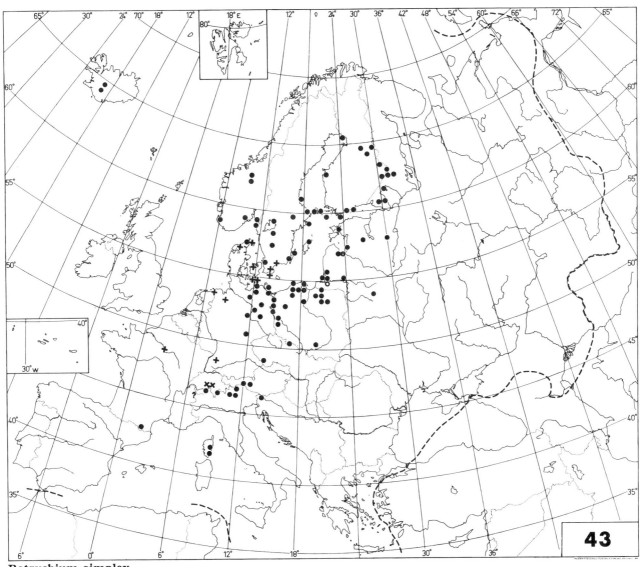

Botrychium simplex

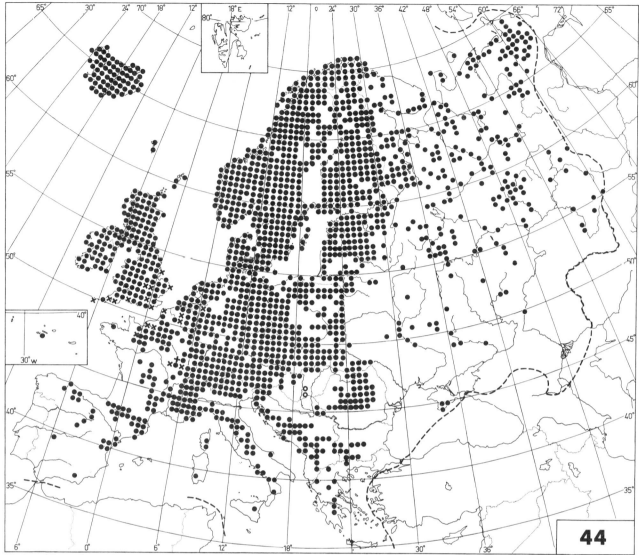

Botrychium lunaria

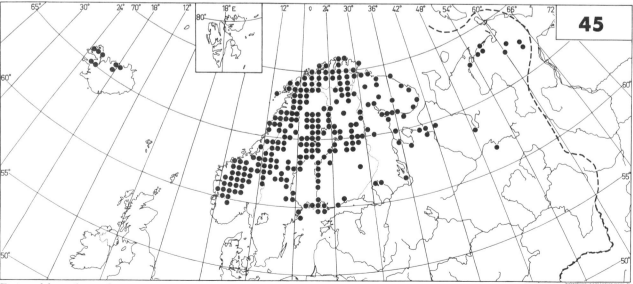

Botrychium boreale

Botrychium lunaria (L.) Swartz — Map 44.

Notes. Sa added (not given in Fl. Eur.).
Total range. Hultén CP 1964: map 103; MJW 1965: map 10b.

Botrychium boreale Milde — Map 45.

Total range. E. Hultén, Fl. Alaska, p. 40 (1968).

Botrychium matricariifolium A. Braun ex Koch — Map 46.

Botrychium ramosum Ascherson, pro parte

Notes. Co and Rs(W) added (not given in Fl. Eur.), in Ho extinct (given as present in Fl. Eur.). In Rm very doubtful (given as certain in Fl. Eur.), all herbarium specimens belonging to *B. multifidum.*
Total range. Hultén AA 1958: map 46; MJW 1965: map 10c.

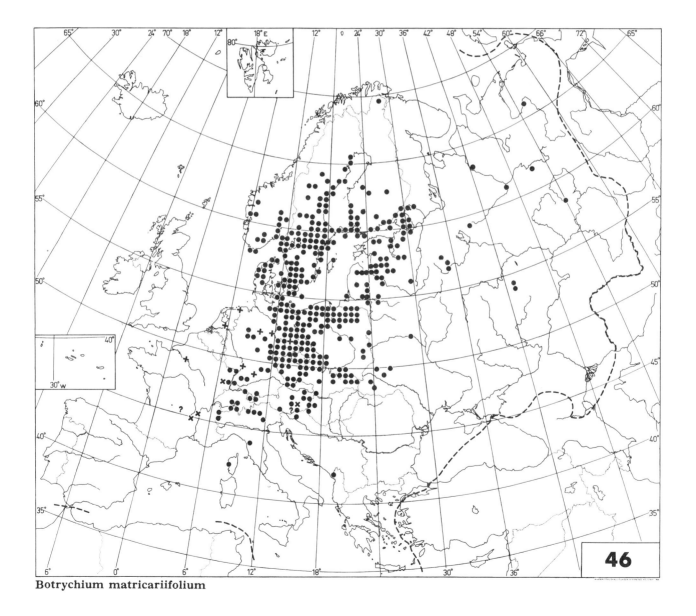

Botrychium matricariifolium

Botrychium lanceolatum (S. G. Gmelin) Ångström — Map 47.

Notes. In Ga probably extinct (given as present in Fl. Eur.), Hs omitted (given in Fl. Eur.), Rs(B) added (not given in Fl. Eur.).

Total range. Hultén AA 1958: map 237; MJW 1965: map 10d.

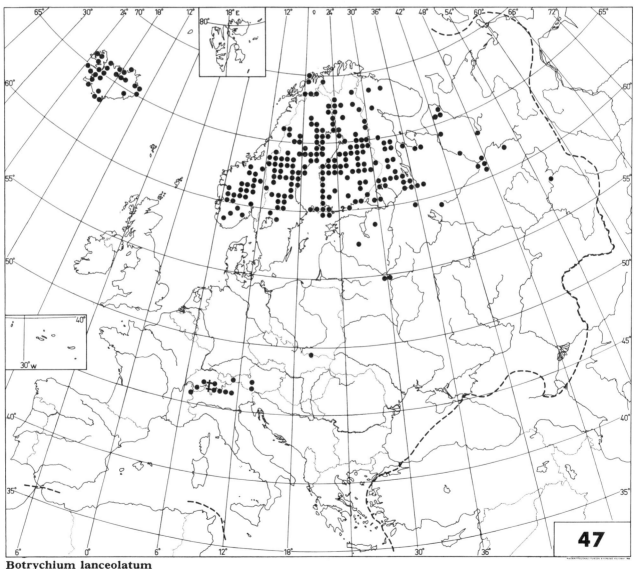

Botrychium lanceolatum

Botrychium multifidum (S. G. Gmelin) Rupr. — Map 48.

Notes. Br omitted (given in Hultén AA 1958, and MJW 1965), Rs(W) added (not given in Fl. Eur.).
Total range. Hultén AA 1958: map 253; MJW 1965: map 11b.

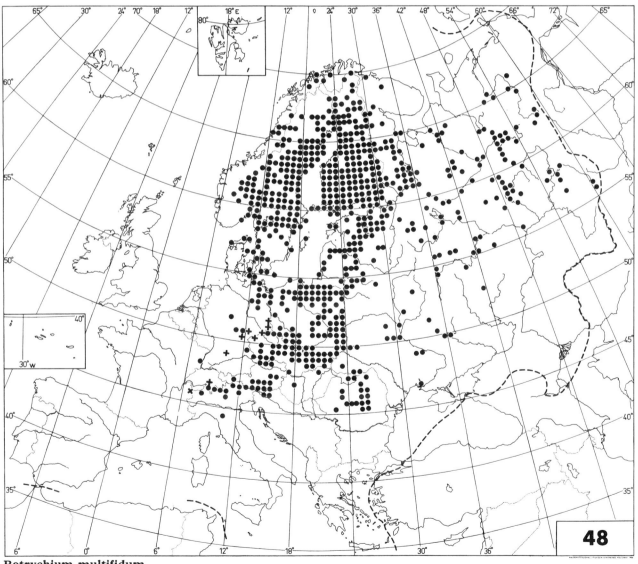

Botrychium multifidum

Botrychium virginianum (L.) Swartz — Map 49.

Notes. In Hu extinct (given as present in Fl. Eur.).
Total range. Hultén CP 1964: map 179.

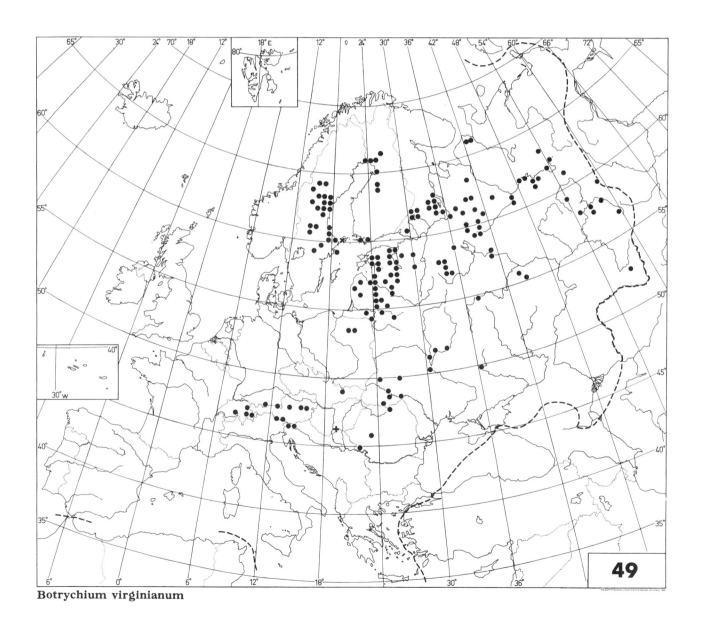

Botrychium virginianum

OSMUNDACEAE

Osmunda regalis L. — Map 50.

Notes. Bl omitted (given in Hultén AA 1958, and MJW 1965), in Cz extinct as native, and naturalized (given as native in Fl. Eur.).

Total range. Hultén AA 1958: map 244; MJW 1965: map 11c; R. E. G. Pichi-Sermolli, Lav. Soc. Ital. Biogeogr. 1: 92 (1971).

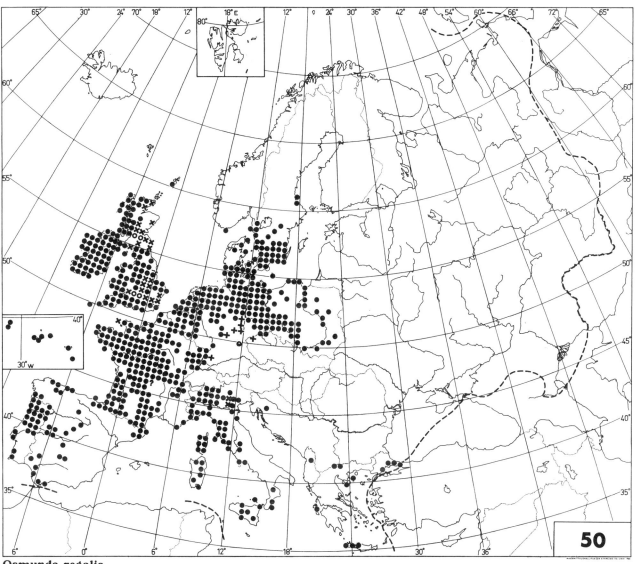

Osmunda regalis

SINOPTERIDACEAE

Cheilanthes marantae (L.) Domin — Map 51.

Notholaena marantae (L.) Desv.

Taxonomy. From Europe, only subsp. *marantae* is known so far. Another diploid (2n = 58), subsp. *subcordata* (Cav.) Benl & Poelt, occurs in Islas Canarias and Madeira. See G. Vida et al., Bauhinia 4: 223—253 (1970).
Notes. Bu and Cz added (not given in Fl. Eur.), Si omitted (given in Fl. Eur.).
Total range. R. E. G. Pichi-Sermolli & V. Chiarino-Maspes, Webbia 17: 410 (1963); R. E. G. Pichi-Sermolli, Lav. Soc. Ital. Biogeogr. 1: 102 (1971).

Cheilanthes pteridioides (Reichard) C. Chr. — Map 52.

Cheilanthes fragrans (L. fil.) Swartz, sensu stricto; C. odora Swartz

Nomenclature. W. Greuter, Boissiera 13: 26—28 (1967); R. E. G. Pichi-Sermolli, Webbia 23: 167—168 (1968).
Taxonomy. Allotetraploid (2n = 120), derived from the diploids *C. maderensis* and *C. persica;* see G. Vida et al., Bauhinia 4: 223—253 (1970).
Notes. So far the tetraploid is known with certainty only from Bl and Anatolia; G. Vida et al., Bauhinia 4: 226 (1970).

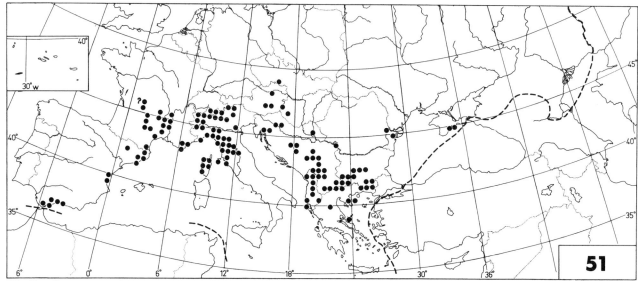

Cheilanthes marantae

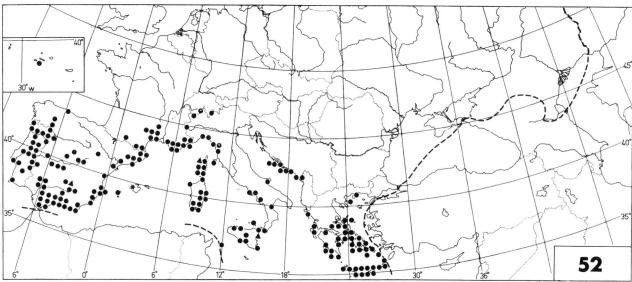

● = **Cheilanthes pteridioides** incl. **C. maderensis**　　★ = **C. pteridioides** sensu stricto
▲ = **C. maderensis**

Cheilanthes maderensis Lowe   Map 52.

Cheilanthes fragrans (L. fil.) Swartz subsp. maderensis (Lowe) Benl

Taxonomy. Diploid (2n = 60); G. Vida et al., Bauhinia 4: 223—253 (1970).

Notes. Known so far from Co, Hs, Si, and Islas Canarias; G. Vida et al., Bauhinia 4: 226—227 (1970).

Cheilanthes hispanica Mett. — Map 53.

Taxonomy. From Hs (Sierra Morena) both a diploid (2n = 60) and a tetraploid (2n = 120) are known. The full identity of the plant of Co (2n = 120) is to be checked. See G. Vida et al., Bauhinia 4: 223—253 (1970).

Notes. Co added (not given in Fl. Eur.).

Total range. Probably endemic to Europe; see G. Vida et al., Bauhinia 4: 231 (1970).

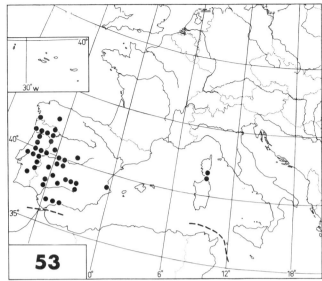

Cheilanthes hispanica

Cheilanthes vellea (Aiton) F. Mueller — Map 54.

Cheilanthes catanensis (Cosent.) H. P. Fuchs; Notholaena lanuginosa (Desf.) Poiret; N. vellea (Aiton) Desv.

Nomenclature. W. Greuter, Boissiera 13: 26—28 (1967).

Taxonomy. The diploid cytotype (2n = 58) is known from Hs, the tetraploid (2n = 116) from Ga, Gr, Hs, Lu, Islas Canarias, and Madeira; G. Vida et al., Bauhinia 4: 223—253 (1970).

Notes. Ga added (not given in Fl. Eur.).

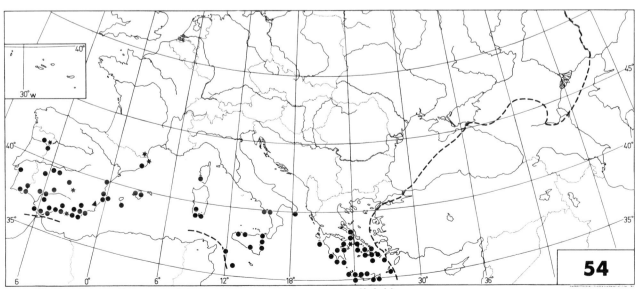

● = **Cheilanthes vellea** sensu lato   ▲ = diploid
                                             ✱ = tetraploid

54

Cheilanthes persica (Bory) Mett. ex Kuhn   Map 55.

    Notes. Probably extinct in It (given as present in Fl. Eur.).

Pellaea calomelanos (Swartz) Link — Map 56.

Pellaea viridis (Forsskål) Prantl — Map 57.

    Notes. [Az] added (not given in Fl. Eur.).
    Native of Africa.

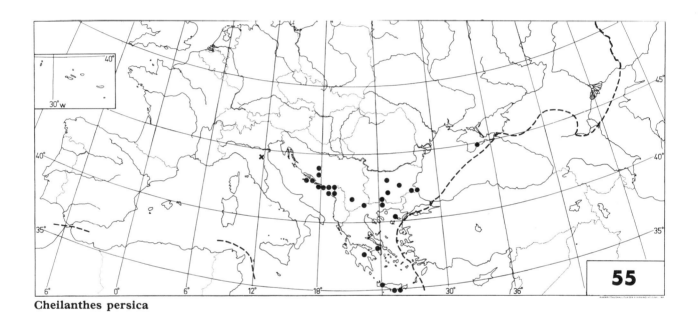

Cheilanthes persica

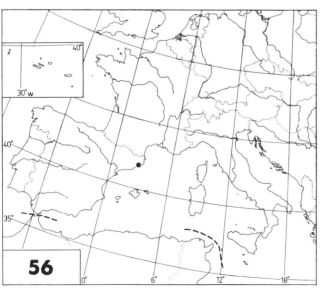

Pellaea calomelanos

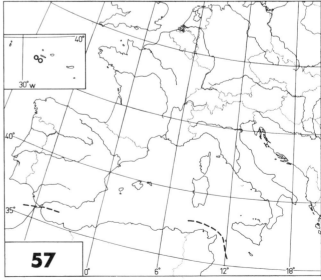

Pellaea viridis

ADIANTACEAE

Adiantum capillus-veneris L. — Map 58.

Notes. Az added (not given in Hultén CP 1964), in Be probably extinct, and Ho omitted ([Be, Ho] in Fl. Eur.), [Hu] added (not given in Fl. Eur.).

Total range. Hultén CP 1964: map 139.

Adiantum raddianum C. Presl

Notes. Not mapped, although recorded as naturalized from gardens in Az; C. M. Ward, Brit. Fern Gaz. 10: 120—121 (1970); J. do Amaral Franco (in litt.).

Native of Tropical America.

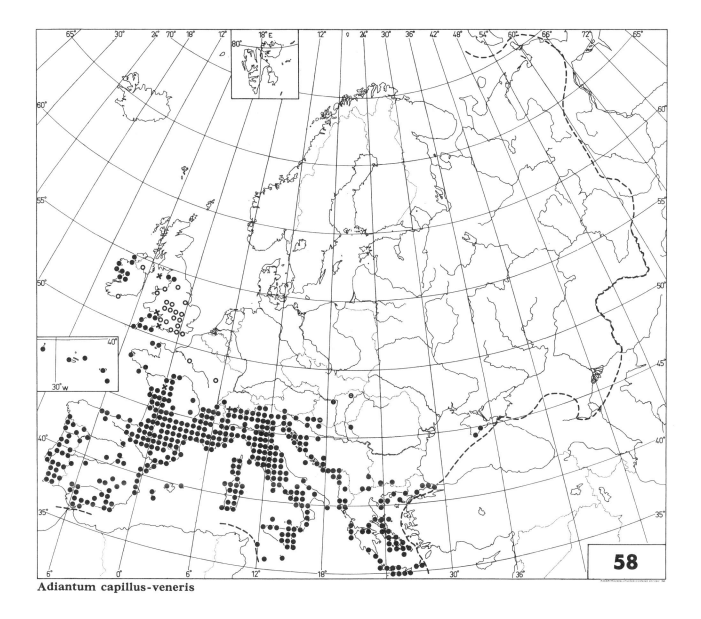

Adiantum capillus-veneris

PTERIDACEAE

Pteris serrulata Forsskål — Map 59.

Pteris arguta Aiton; P. palustris Poiret

Nomenclature. H. Runemark, Bot. Not. 115: 190 (1962); C. V. Morton, Bull. Soc. Bot. France 116: 247—248 (1970).

Notes. Hs added (not given in Fl. Eur.), the plant being recorded for the first time from the European mainland by B. Molesworth Allen, Anal. Univ. Hispal. Ci. 27: 149—151 (1967), Lagascalia 1: 85—86 (1971), Brit. Fern Gaz. 10: 200—201 (1971).

Total range. H. Runemark, Bot. Not. 115: 184 (1962).

Pteris cretica L. — Map 59.

Notes. Az, Ho and Hu omitted ([Az, Ho, Hu] in Fl. Eur.), Cr omitted (given in Fl. Eur.), in Hs native ([Hs] in Fl. Eur.), in Sa perhaps not native (given as native in Fl. Eur.).

Pteris vittata L. — Map 60.

Notes. Bl omitted (given in Fl. Eur.), [Ga] added (not given in Fl. Eur.).

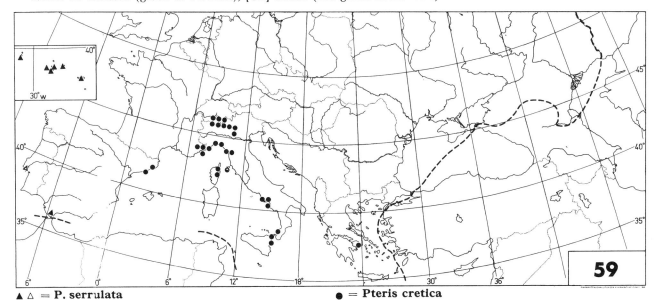

▲ △ = P. serrulata   ● = Pteris cretica

59

Pteris vittata

60

CRYPTOGRAMMACEAE

Cryptogramma crispa (L.) Hooker — Map 61.

Allosorus crispus (L.) Röhling

Notes. Fa and Rs(C) omitted (given in Fl. Eur.).
Total range. Hultén AA 1958: map 225; MJW 1965: map 12b.

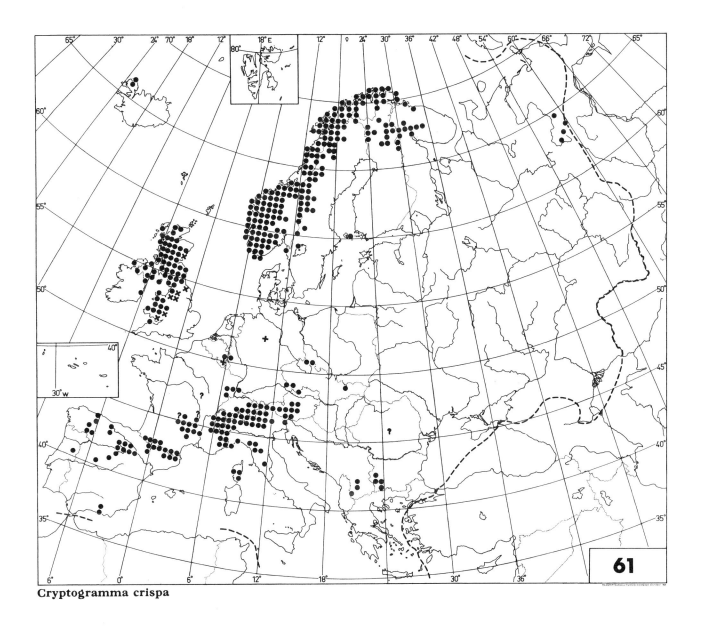

Cryptogramma crispa

Cryptogramma stelleri (S. G. Gmelin) Prantl — Map 62.

Total range. E. Hultén, Fl. Alaska, p. 45 (1968).

Onychium japonicum (Thunb.) G. Kunze — Map 63.

Notes. [Az] added (not given in Fl. Eur.).
Native of east Asia.

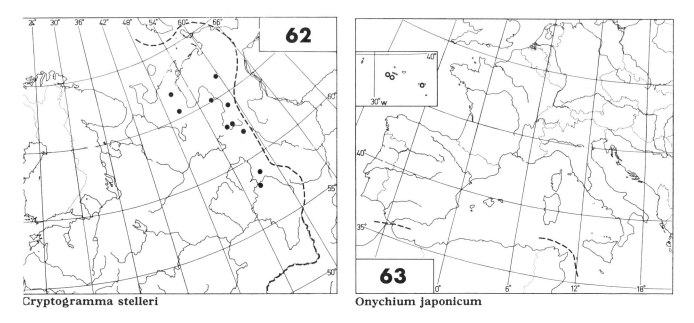

Cryptogramma stelleri

Onychium japonicum

HEMIONITIDACEAE (GYMNOGRAMMACEAE)

Anogramma leptophylla (L.) Link — Map 64.

Gymnogramma leptophylla (L.) Desv.

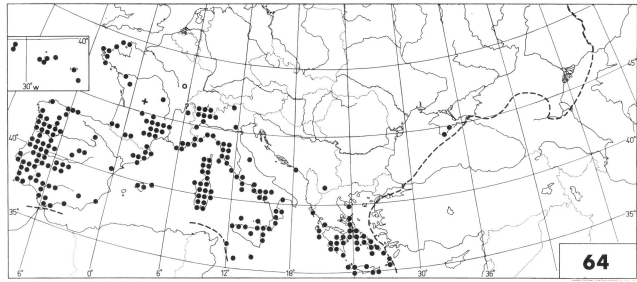

Anogramma leptophylla

Pityrogramma chrysophylla (Swartz) Link —
Map 65.

Notes. [Az] added (not given in Fl. Eur.).
Native of Africa.

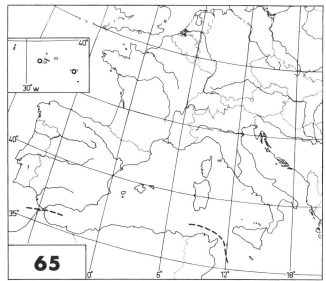

65

Pityrogramma chrysophylla

DICKSONIACEAE

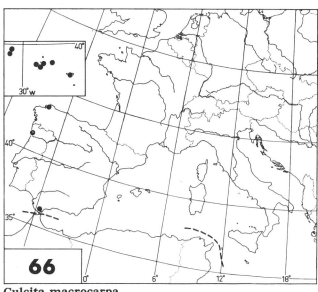

66

Culcita macrocarpa

Culcita macrocarpa C. Presl — Map 66.

Notes. Obviously introduced in Lu (given as
native in Fl. Eur.).
Total range. R. E. Holttum, Proc. Roy. Soc.
London (B) 161: 43 (1965).

HYPOLEPIDACEAE

Pteridium aquilinum (L.) Kuhn

Pteris aquilina L.

Total range. Hultén CP 1964: map 131; MJW 1965: map 12a.

P. aquilinum subsp. **aquilinum** — Map 67.

Taxonomy. Widespread and polymorphic. Within Europe two taxa overlap: var. *typicum* sensu Tryon (? = var. *aquilinum*) in the S, SW and W, and var. *latiusculum* Underw. in the N. They cannot be mapped separately here. A deviating chromosome number (2n = 52, instead of 2n = 104) was recently reported for material from calcareous rocks of southern Hs by B. Molesworth Allen, Brit. Fern Gaz. 10: 34—36 (1968), and Á. & D. Löve, V Simp. Fl. Eur. Trab. Comun., p. 290 (Sevilla 1969).

P. aquilinum subsp. **brevipes** (Tausch) Wulf — Map 67.

Notes. Cr added (not given in Fl. Eur.). Found in Kriti in 1971 by A. C. Jermy (unpubl.).

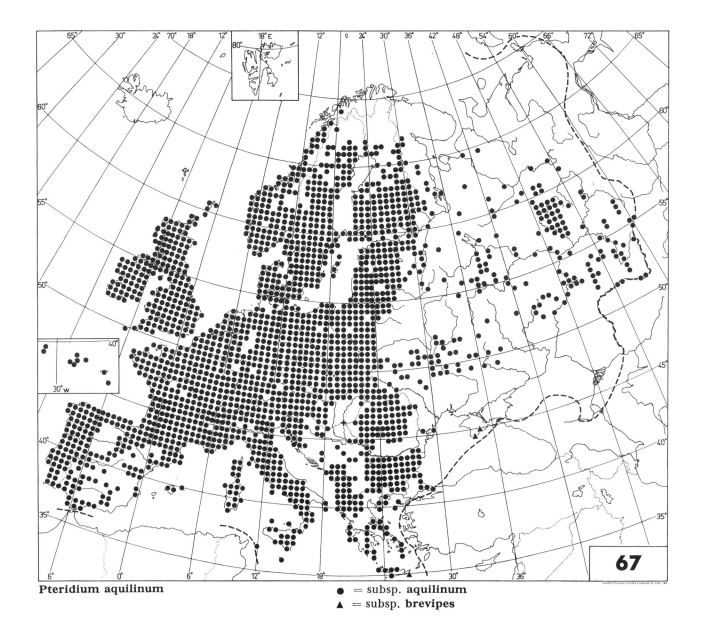

Pteridium aquilinum

● = subsp. **aquilinum**
▲ = subsp. **brevipes**

DAVALLIACEAE

Davallia canariensis (L.) Sm. — Map 68.

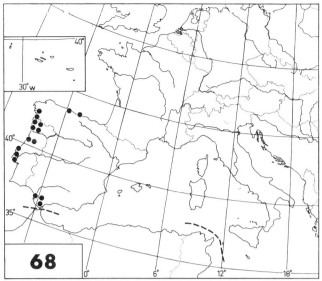

68

Davallia canariensis

HYMENOPHYLLACEAE

69

Hymenophyllum tunbrigense

Hymenophyllum tunbrigense (L.) Sm. — Map 69.

Notes. Co omitted (?Co in Fl. Eur.), Cz added as probably extinct (not given in Fl. Eur.).

Total range. MJW 1965: map 11d; R. E. G. Pichi-Sermolli, Lav. Soc. Ital. Biogeogr. 1: 94 (1971).

62

Hymenophyllum wilsonii Hooker — Map 70.

Hymenophyllum peltatum auct., vix Desv.; H. unilaterale auct., vix Bory

Total range. Hultén CP 1964: map 148.

Trichomanes speciosum Willd. — Map 71.

Trichomanes radicans auct., vix Swartz

Notes. In Lu partly extinct, partly introduced (given as present and native in Fl. Eur.).
Total range. R. A. Fataliyev, Bot. Žur. 45: 1215 (1960).

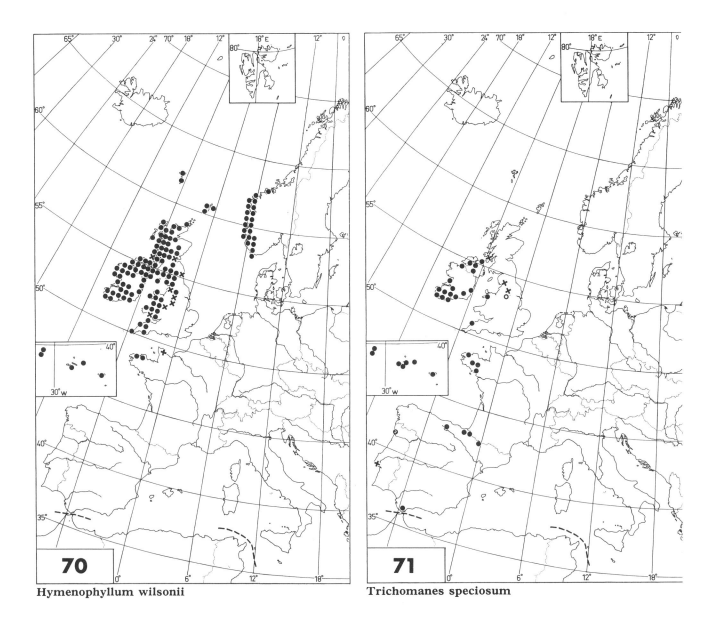

Hymenophyllum wilsonii

Trichomanes speciosum

THELYPTERIDACEAE

Thelypteris limbosperma (All.) H. P. Fuchs — Map 72.

Dryopteris oreopteris (Ehrh.) Maxon; Lastrea oreopteris (Ehrh.) Bory; Nephrodium oreopteris (Ehrh.) Desv.; Oreopteris limbosperma (All.) J. Holub; Polystichum oreopteris (Ehrh.) DC.

Nomenclature. J. Holub, Folia Geobot. Phytotax. 4: 33—53 (1969).

Notes. Az added (not given in Fl. Eur.).

Total range. Hultén CP 1964: map 144; MJW 1965: map 17a; R. E. G. Pichi-Sermolli, Lav. Soc. Ital. Biogeogr. 1: 117 (1971).

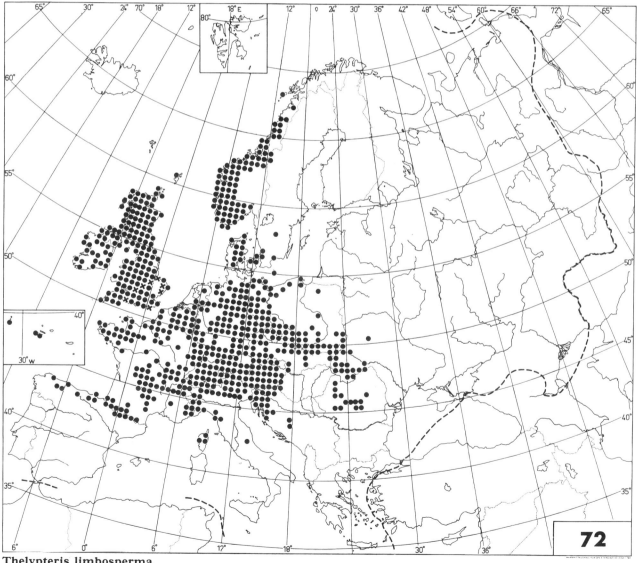

Thelypteris limbosperma

64

Thelypteris palustris Schott — Map 73.

Dryopteris thelypteris (L.) A. Gray; Lastrea thelypteris (L.) Bory; Nephrodium thelypteris (L.) Strempel; Polystichum thelypteris (L.) Roth

Notes. Rs(E) confirmed (?Rs(E) in Fl. Eur.).
Total range. Hultén CP 1964: map 170; R. E. G. Pichi-Sermolli, Lav. Soc. Ital. Biogeogr. 1: 116 (1971).

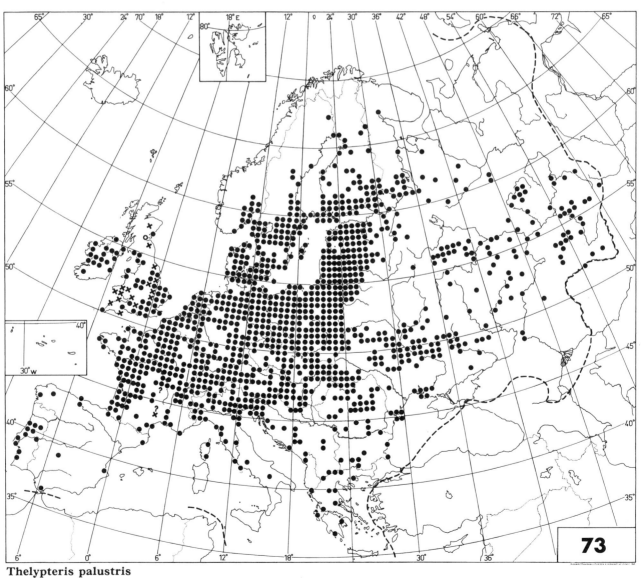

Thelypteris palustris

Thelypteris phegopteris (L.) Slosson — Map 74.

Dryopteris phegopteris (L.) C. Chr.; Lastrea phegopteris (L.) Bory; Nephrodium phegopteris (L.) Prantl; Phegopteris connectilis (Michx) Watt.; P. polypodioides Fée; P. vulgaris Mett.; Polypodium phegopteris L.

Nomenclature. Should perhaps more properly be regarded as belonging to a separate genus *Phegopteris*, as *P. connectilis* (Michx) Watt.; see R. E. Holttum, Blumea 19: 26 (1971).

Notes. Rs(E) added (not given in Fl. Eur.).

Total range. Hultén CP 1964: map 107; MJW 1965: map 16d.

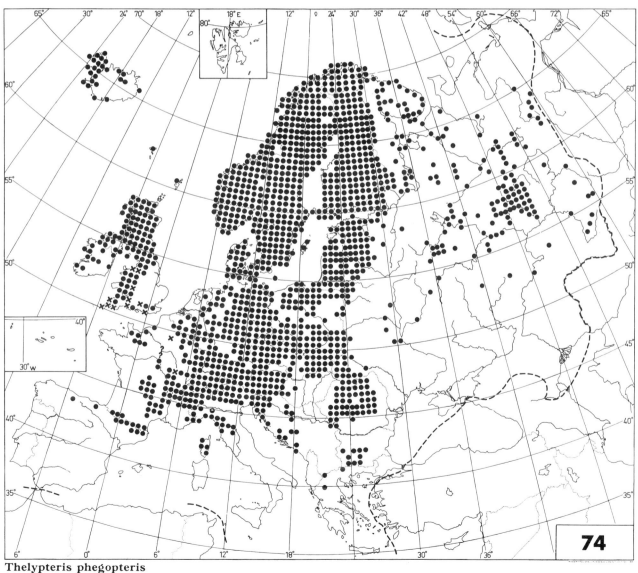

Thelypteris phegopteris

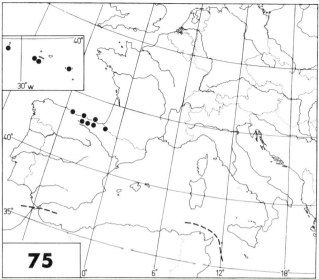

Thelypteris pozoi

Thelypteris pozoi (Lag.) C. V. Morton — Map 75.

Hemionitis pozoi Lag.; Leptogramma pozoi (Lag.) Heywood; Stegnogramma pozoi (Lag.) K. Iwatsuki; Ceterach hispanicum sensu Willk., pro parte.

Taxonomy and nomenclature. If the genus *Thelypteris* is divided, as recently proposed by several specialists, *T. pozoi* should be placed in *Stegnogramma* Blume; see K. Iwatsuki, Acta Phytotax. Geobot. 19: 112—126 (1963), R. E. Holttum, Blumea 19: 17—52 (1971).

Notes. Ga added (not given in Fl. Eur.).

Cyclosorus dentatus (Forsskål) R.-C. Ching — Map 76.

Aspidium molle Swartz

Taxonomy and nomenclature. Obviously to be placed in the genus *Christella* Léveillé; see R. E. Holttum, Blumea 19: 17—52 (1971).

Notes. Cr added (not given in Fl. Eur.), found in 1971 by A. C. Jermy (unpubl.); Hs added (not given in Fl. Eur.), this being the first record for the European mainland, see B. Molesworth Allen, Bol. Soc. Esp. Hist. Nat. 67: 75—76 (1969), Brit. Fern Gaz. 10: 202 (1971).

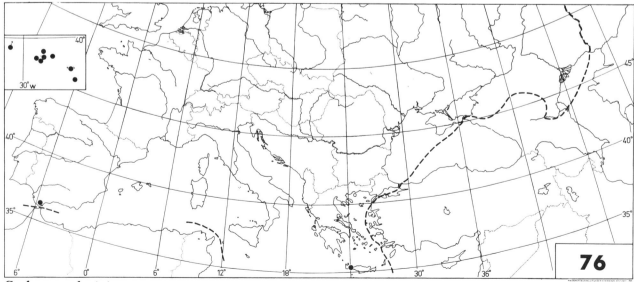

Cyclosorus dentatus

ASPLENIACEAE

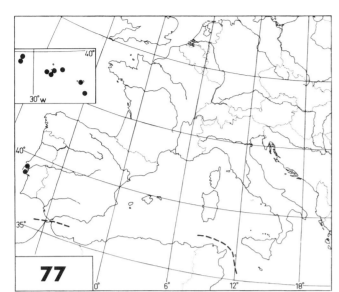

77

Asplenium hemionitis

Asplenium hemionitis L. — Map 77.

Asplenium palmatum Lam.

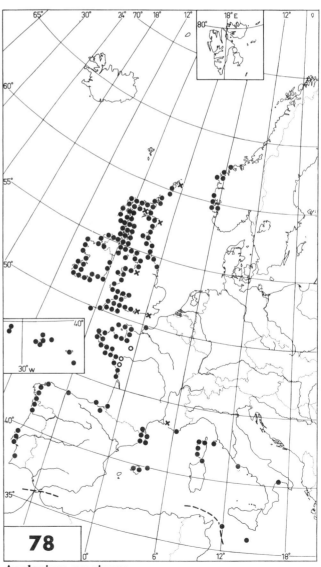

78

Asplenium marinum

Asplenium marinum L. — Map 78.

Asplenium petrarchae (Guérin) DC.

Asplenium glandulosum Lois.

Notes. Sa added (not given in Fl. Eur.).
Total range. Endemic to Europe (Fl. Eur.). However, the map by P. V. Arrigoni & C. Ricceri, Webbia 24: 420 (1969), gives several localities in N. Africa.

A. petrarchae subsp. **petrarchae** — Map 79.

Taxonomy. Tetraploid.
Notes. In Al uncertain (given as certain in Fl. Eur.).

A. petrarchae subsp. **bivalens** (D. E. Meyer) Lovis & Reichstein — Map 79.

Asplenium glandulosum subsp. bivalens D. E. Meyer

Taxonomy and nomenclature. Diploid. D. E. Meyer, Ber. Deutsch. Bot. Ges. 77: 7 (1964); J. D. Lovis & T. Reichstein, Ber. Schweiz. Bot. Ges. 79: 336 (1969).

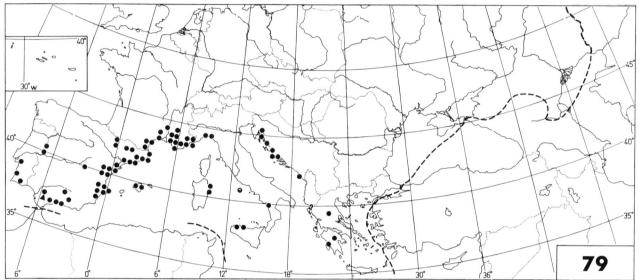

Asplenium petrarchae
● = subsp. **petrarchae**
▲ = subsp. **bivalens** & subsp. **petrarchae**

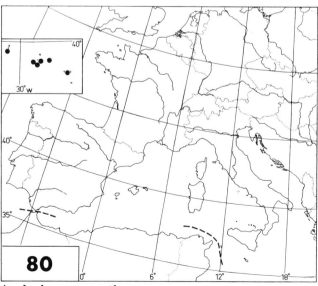

Asplenium monanthes L. — Map 80.

Asplenium monanthes

Asplenium trichomanes L. — Map 81.

Taxonomy. Three subspecies (J. D. Lovis, Brit. Fern Gaz. 9: 147—160 (1964)). Present data do not allow their mapping as separate units.

Total range. Hultén CP 1964: map 130; MJW 1965: map 13a.

A. trichomanes subsp. trichomanes

Diploid. Mainly on non-calcareous rocks. Throughout the range, more in the north and upland regions.

A. trichomanes subsp. quadrivalens D. E. Meyer

Autotetraploid (J. D. Lovis & T. Reichstein, Ber. Schweiz. Bot. Ges. 79: 341 (1969)). Largely on basic or calcareous rocks. Throughout the range.

A. trichomanes subsp. inexpectans Lovis

Diploid. Basic or calcareous rocks. As far as is known, endemic to Europe (Au, Gr, Ju).

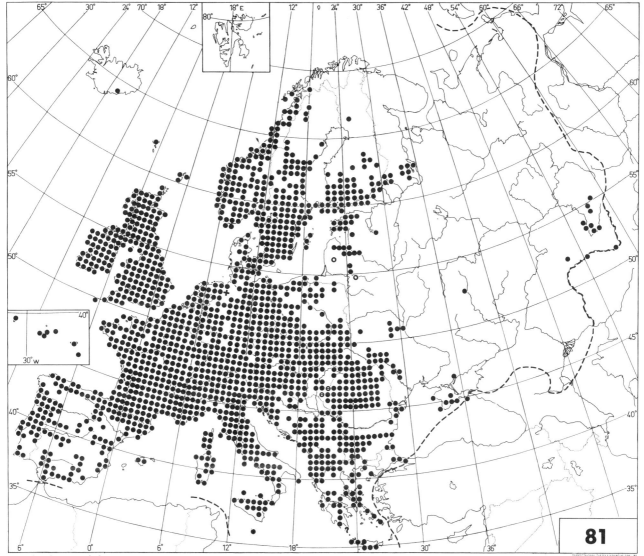

Asplenium trichomanes

Asplenium adulterinum Milde — Map 82.

Taxonomy. A tetraploid derivative of the hybrid *A. trichomanes* subsp. *trichomanes* × *A. viride*, perhaps arisen several times; see J. D. Lovis & T. Reichstein, Bauhinia 4: 53—63 (1969).

Notes. Ga omitted (given in MJW 1965), Gr and It added (not given in Fl. Eur.).

Total range. Endemic to Europe.

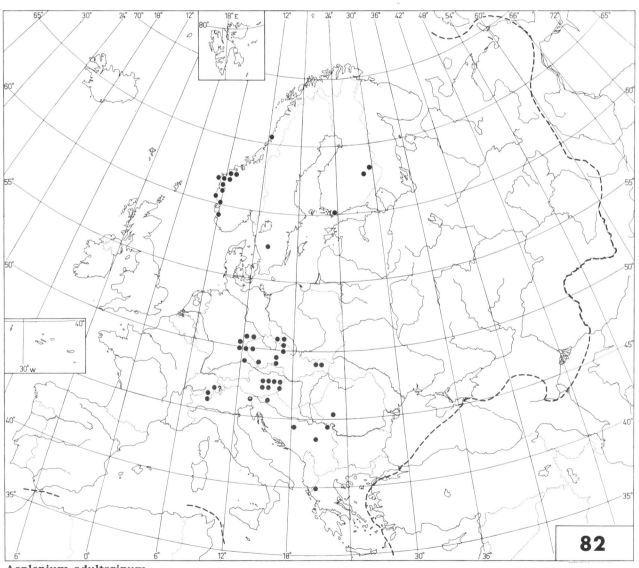

Asplenium adulterinum

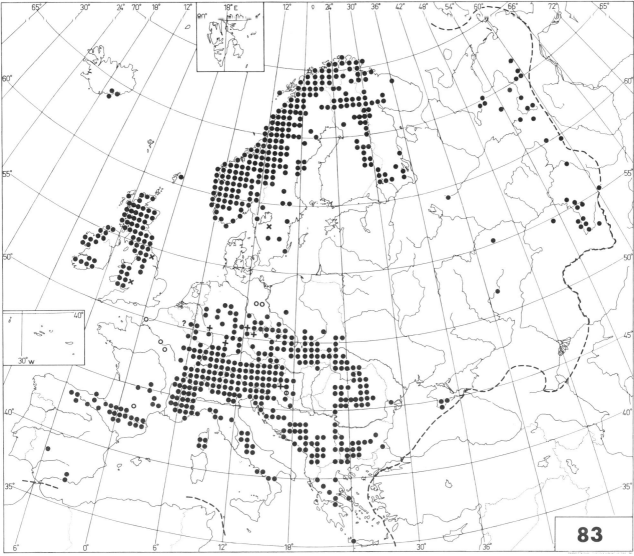

Asplenium viride

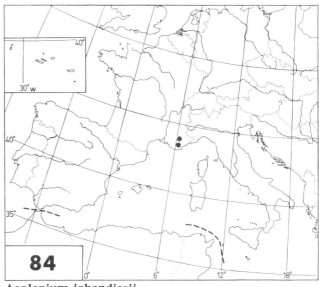

Asplenium jahandiezii

Asplenium viride Hudson — Map 83.

Notes. Cr and Rs(E) added (not given in Fl. Eur.).

Total range. Hultén CP 1964: map 92; MJW 1965: map 13b.

Asplenium jahandiezii (Litard.) Rouy — Map 84.

Total range. Endemic to Europe.

72

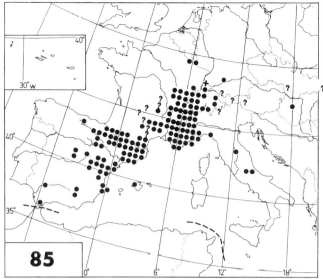

Asplenium fontanum

Asplenium fontanum (L.) Bernh. — Map 85.

Asplenium halleri (Roth) DC.

Notes. †Be added (not given in Fl. Eur.), Co, Cr and Gr omitted (given in Fl. Eur.), in Cz uncertain (given as certain in Fl. Eur.). In central Ga confused with *A. forisiense*.
Total range. MJW 1965: map 13d.

Asplenium majoricum Litard. — Map 86.

Asplenium lanceolatum var. majoricum (Litard.) Sennen & Pau

Taxonomy. A. C. Jermy & J. D. Lovis, Brit. Fern Gaz. 9: 163—167 (1964). Tetraploid, derived from the hybrid *A. fontanum* × *A. petrarchae* subsp. *bivalens*, according to J. D. Lovis & T. Reichstein, Ber. Schweiz. Bot. Ges. 79: 335—345 (1969).
Total range. Endemic to Islas Baleares. Map in C. Jaquotot & J. Orell, Collect. Bot. 7: 561 (1968).

Asplenium forisiense Le Grand — Map 87.

Asplenium foresiacum (Le Grand) Christ

Notes. In Be found in 1858 and now extinct (?Be in Fl. Eur.), Hs confirmed (?Hs in Fl. Eur.). The occurrence in Co and Sa doubted by T. Reichstein (in litt.).
Total range. Endemic to Europe.

Asplenium macedonicum Kümmerle — Map 87.

Incl. Asplenium bornmuelleri Kümmerle

Total range. Endemic to Europe.

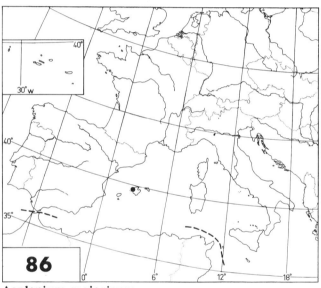

Asplenium majoricum

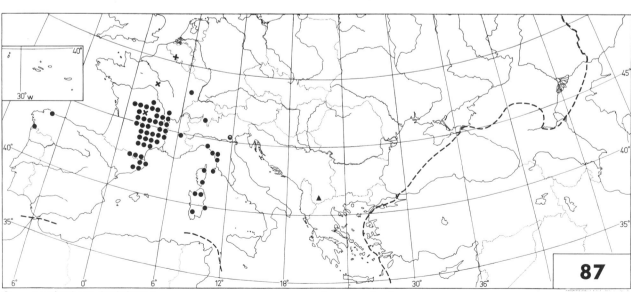

● = **Asplenium forisiense**   ▲ = **A. macedonicum**

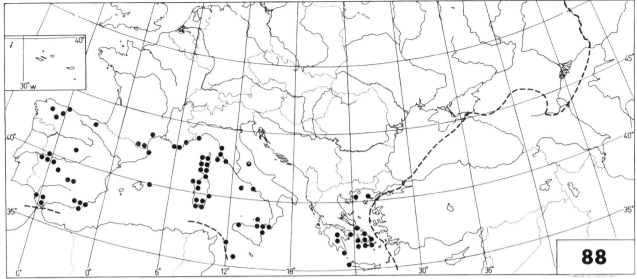

Asplenium obovatum

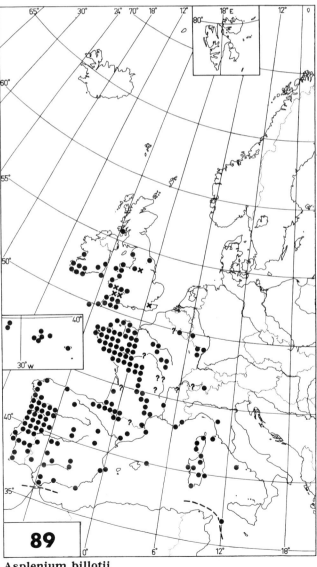

Asplenium billotii

Asplenium obovatum Viv. — Map 88.

Asplenium lanceolatum auct., non Hudson

Notes. Bl confirmed (?Bl in Fl. Eur.), Cr and ?Ju added (not given in Fl. Eur.), Lu and Tu omitted (given in Fl. Eur.). The records from northern and central Spain perhaps refer to *A. billotii.*

Asplenium balearicum Shivas

A tetraploid derivative from the cross *A. obovatum* × *A. onopteris;* see M. G. Shivas, Brit. Fern Gaz. 10: 68—80 (1969). Not mapped. As far as is known, endemic to Islas Baleares.

Asplenium billotii F. W. Schultz — Map 89.

Asplenium lanceolatum Hudson, non Forsskål

Notes. Ge omitted (given in Fl. Eur.), He added (not given in Fl. Eur.).

Asplenium adiantum-nigrum L. — Map 90.

Notes. Az omitted (given in Fl. Eur.), Bl added (not given in Fl. Eur.), in Cr doubtful (given as certain in Fl. Eur.).

Total range. Hultén CP 1964: map 142.

Asplenium onopteris L. — Map 91.

Notes. ?Be and He added (not given in Fl. Eur.), Br and ?Ge omitted (given in Fl. Eur.). The plants from Be and from western and central Ga are reported as being not quite typical.

Total range. Approximative northern and eastern limit in Hultén CP 1964: map 142.

Asplenium cuneifolium Viv. — Map 92.

Asplenium serpentini Tausch; incl. A. forsteri Sadler & A. silesiacum Milde

Notes. Bu and Rs(W) added (not given in Fl. Eur.), Co omitted (given in Fl. Eur.), Hs confirmed (?Hs in Fl. Eur.).

Total range. MJW 1965: map 14a.

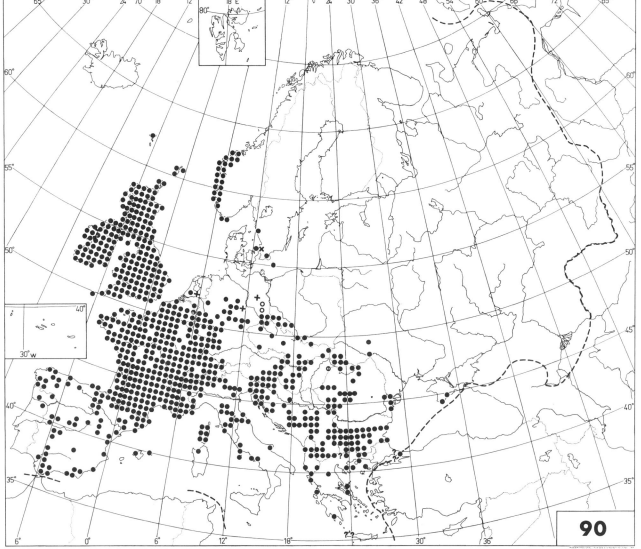

Asplenium adiantum-nigrum

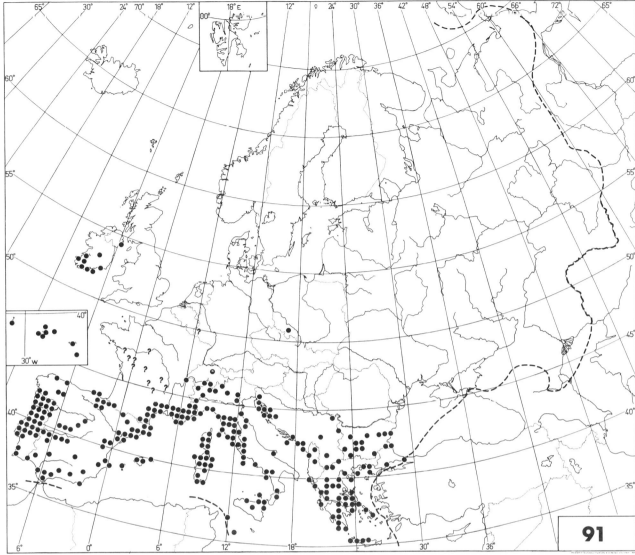

Asplenium onopteris

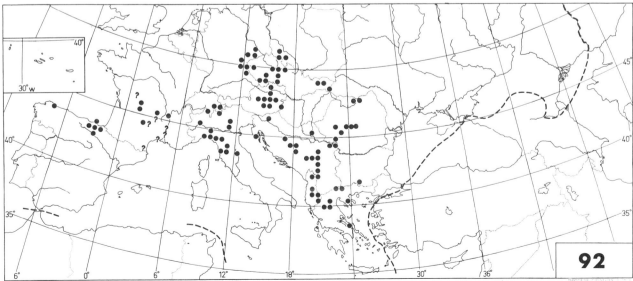

Asplenium cuneifolium

Asplenium septentrionale (L.) Hoffm. — Map 93.

Notes. Hb, Is and Rs(B, E) added (not given in Fl. Eur.).
Total range. Hultén AA 1958: map 228; MJW 1965: map 14b.

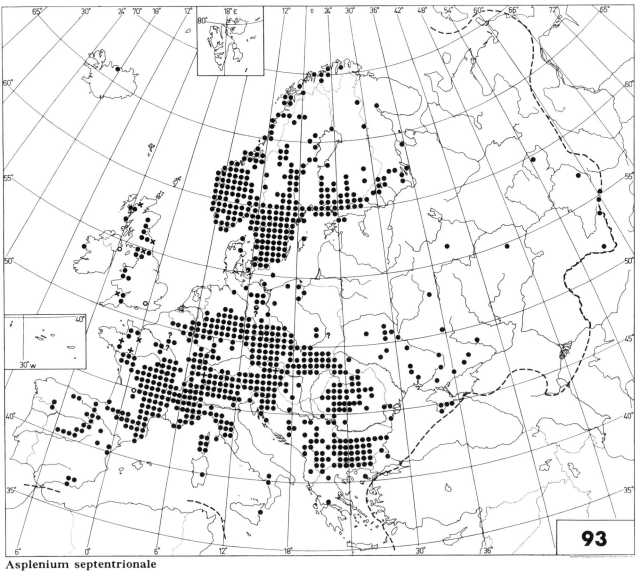

Asplenium septentrionale

Asplenium seelosii Leybold

A. seelosii subsp. **seelosii** — Map 94.

>*Taxonomy.* Diploid.
>*Total range.* Endemic to Europe.

A. seelosii subsp. **glabrum** (Litard. & Maire) Rothm. — Map 94.

Asplenium celtibericum Rivas-Martínez

>*Taxonomy and nomenclature.* Diploid (D. E. Meyer, Ber. Deutsch. Bot. Ges. 80: 37 (1967)). S. Rivas-Martínez, Bull. Jard. Bot. Nat. Belg. 37: 329 (1967).
>*Notes.* Ga added (not given in Fl. Eur.).
>*Total range.* Endemic to Europe (Fl. Eur.). However, one of the syntypes of subsp. *glabrum* is from Morocco; see R. Maire, Bull. Soc. Sci. Nat. Maroc 8: 143 (1929) (also in MJW 1965: map 14c).

Asplenium eberlei D. E. Meyer — Map 95.

>*Taxonomy.* Originally described as a hybrid by D. E. Meyer, Ber. Deutsch. Bot. Ges. 75: 29 (1962). It has proved to be a fertile allotetraploid, probably derived from hybridization between *A. ruta-muraria* subsp. *dolomiticum* and *A. seelosii* subsp. *seelosii*; D. E. Meyer, Ber. Deutsch. Bot. Ges. 80: 28—32 (1967).
>*Total range.* Endemic to Europe.

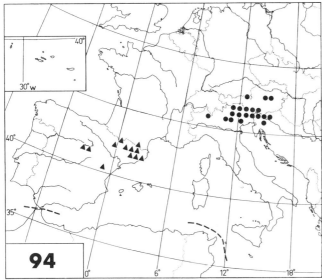

Asplenium seelosii
● = subsp. **seelosii**
▲ = subsp. **glabrum**

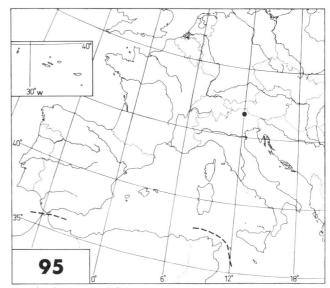

Asplenium eberlei

Asplenium ruta-muraria L.

Notes. Cr added (not given in Fl. Eur.).
Total range. Hultén CP 1964: map 156.

A. ruta-muraria subsp. **ruta-muraria** — Map 96.

Taxonomy. Autotetraploid, most probably derived from subsp. *dolomiticum;* G. Vida, Caryologia 23: 545 (1970).

A. ruta-muraria subsp. **dolomiticum** Lovis & Reichstein — Map 96.

Taxonomy and nomenclature. Diploid. J. D. Lovis & T. Reichstein, Brit. Fern Gaz. 9: 143 (1964).
Total range. Endemic to Europe.

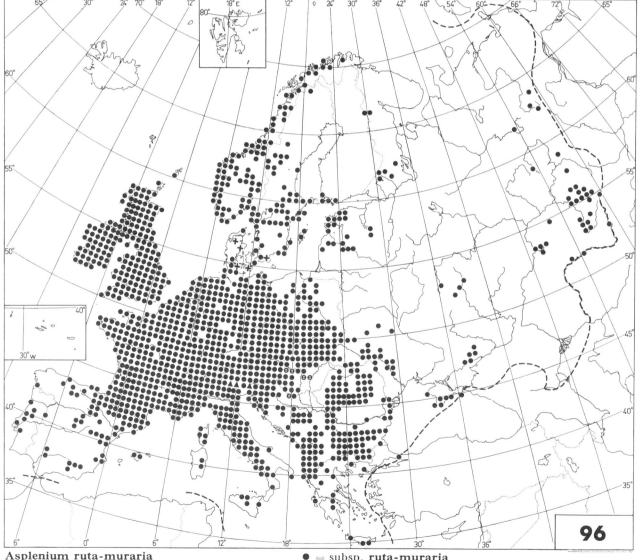

Asplenium ruta-muraria

● = subsp. **ruta-muraria**
▲ = subsp. **dolomiticum** & subsp. **ruta-muraria**

Asplenium lepidum C. Presl — Map 97.

Notes. Cr confirmed (?Cr in Fl. Eur.).

Asplenium haussknechtii Godet & Reuter — Map 97.

Taxonomy. This taxon, described from Anatolia, should perhaps more properly be classified as a subspecies under *A. lepidum.* In addition, two undescribed species of the same affinity have recently been discovered in Kriti, according to information given by T. Reichstein.
Notes. Cr added (the species not mentioned in Fl. Eur.). The records are based solely on material checked by T. Reichstein.

Asplenium fissum Kit. ex Willd. — Map 98.

Notes. Ga confirmed (?Ga in Fl. Eur.), He added (not given in Fl. Eur.).
Total range. Endemic to Europe.

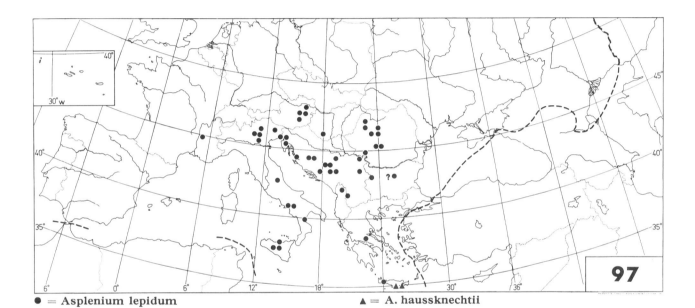

● = Asplenium lepidum   ▲ = A. haussknechtii

Asplenium fissum

80

Ceterach officinarum DC.

Asplenium ceterach L.

Notes. In Au an established alien (given as native in Fl. Eur.).
Total range. MJW 1965: map 14d; R. E. G. Pichi-Sermolli, Lav. Soc. Ital. Biogeogr. 1: 105 (1971).

C. officinarum subsp. officinarum — Map 99.

Taxonomy. Tetraploid.
Notes. The map includes the collective records.

C. officinarum subsp. bivalens D. E. Meyer — Map 100.

Asplenium javorkeanum Vida; Ceterach javorkeanum (Vida) Soó

Taxonomy and nomenclature. Diploid. D. E. Meyer, Ber. Deutsch. Bot. Ges. 77: 8 (1964).
Notes. See under subsp. *officinarum*.
Total range. As far as is known, endemic to Europe; cf. G. Vida, Acta Bot. Acad. Sci. Hung. 9: 206—207 (1963).

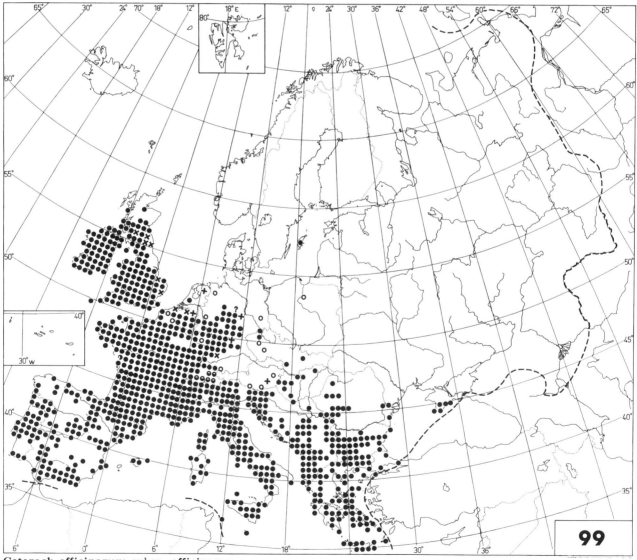

99

Ceterach officinarum subsp. officinarum

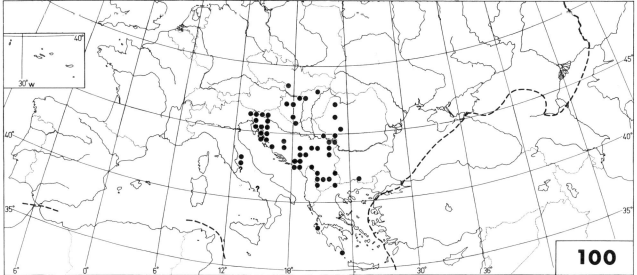

Ceterach officinarum subsp. **bivalens**

Pleurosorus hispanicus (Cosson) C. V. Morton
— Map 101.

Ceterach hispanicum (Cosson) Mett.

Notes. Contrary to the statement in Fl. Eur., no records have been obtained from northwestern Hs.

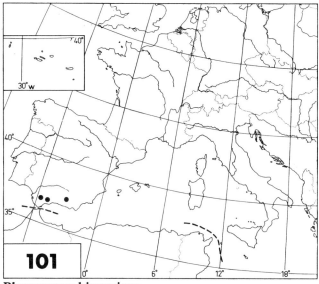

Pleurosorus hispanicus

Phyllitis scolopendrium (L.) Newman — Map 102.

Asplenium scolopendrium L.; Scolopendrium officinale Sm.; incl. Biropteris antri-jovis Kümmerle

Notes. The Cretan taxon (*Biropteris antri-jovis*) has been provisionally mapped with this species (in Fl. Eur. under *P. sagittata*).

Total range. Hultén CP 1964: map 147; MJW 1965: map 12d.

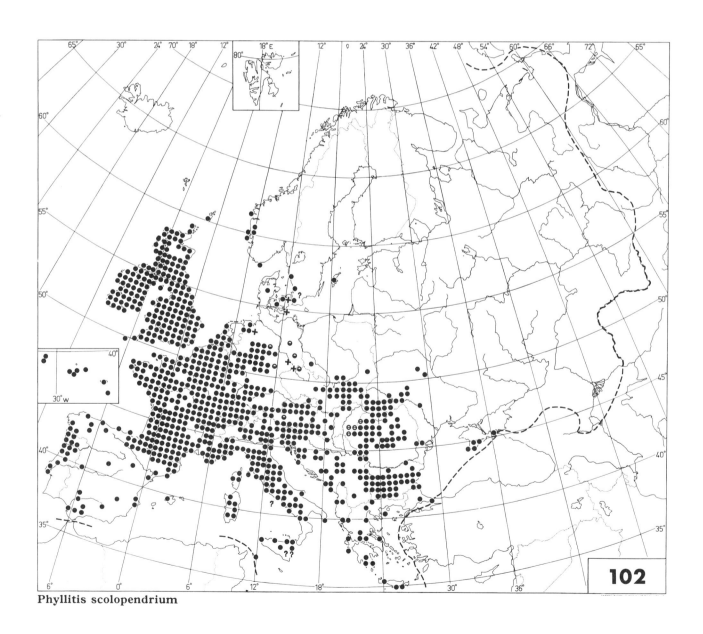

Phyllitis scolopendrium

102

Phyllitis sagittata (DC.) Guinea & Heywood — Map 103.

Phyllitis hemionitis O. Kuntze; Scolopendrium hemionitis Swartz

Notes. In Cr doubtful, see under *P. scolopendrium.*
Total range. F. Morton, Österr. Bot. Zeitschr. 64: 22 (1914); G. Vida, Acta Bot. Acad. Sci. Hung. 9: 207 (1963).

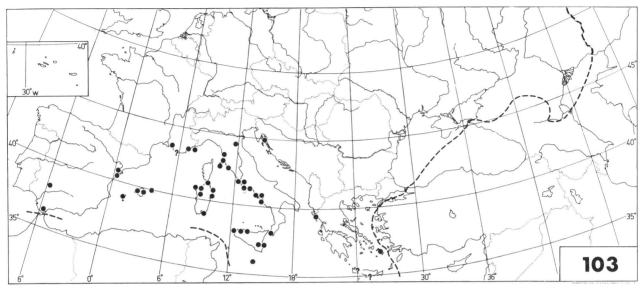

Phyllitis sagittata

Phyllitis hybrida (Milde) C. Chr. — Map 104.

Taxonomy. Tetraploid, perhaps derived from the hybrid *P. sagittata* × *Ceterach officinarum* subsp. *bivalens.*
Total range. Endemic to Europe.

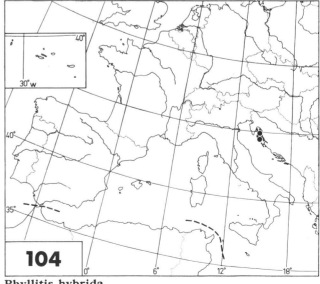

Phyllitis hybrida

ATHYRIACEAE

Athyrium filix-femina (L.) Roth — Map 105.

Notes. Rs(K) omitted (given in Fl. Eur.).
Total range. Hultén CP 1964: map 168; MJW 1965: map 15a.

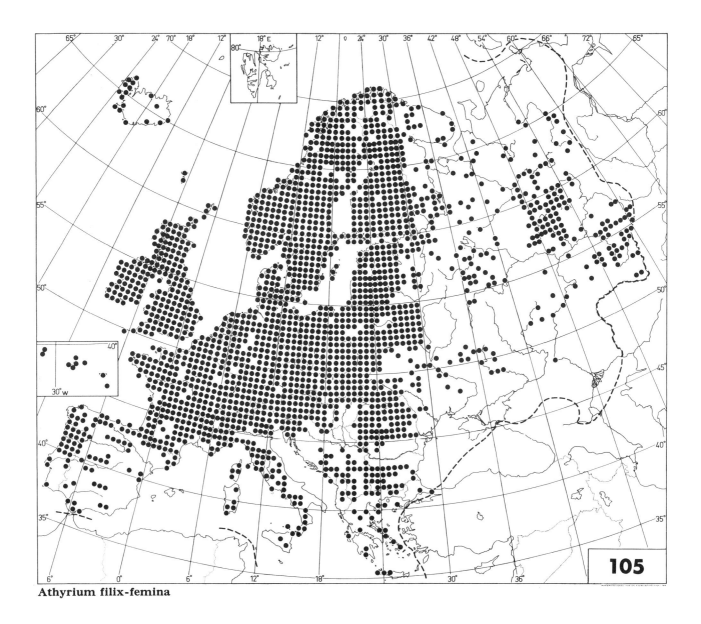

Athyrium filix-femina

Athyrium distentifolium Tausch ex Opiz — Map 106.

Athyrium alpestre (Hoppe) Rylands, non Clairv.

Taxonomy. In Scotland also the endemic var. *flexile* (Newman) Jermy (*A. flexile* (Newman) Druce); see F. H. Perring, Crit. Suppl. Atlas Brit. Fl., p. 2 (1968).

Notes. Da omitted (given in MJW 1965), Fa added (not given in Fl. Eur.), Hb omitted (given in Hultén AA 1958); Rs(C) confirmed (?Rs(C) in Fl. Eur.).

Total range. Hultén AA 1958: map 223; MJW 1965: map 15b.

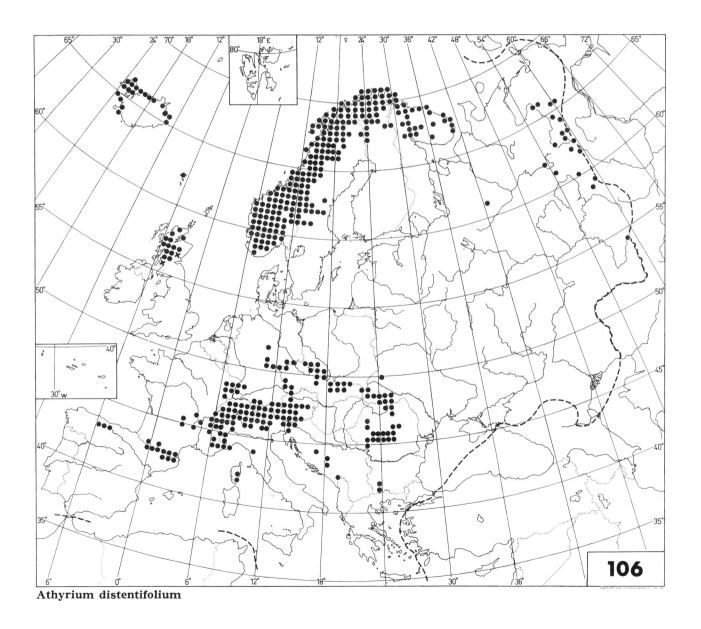

Athyrium distentifolium

Diplazium caudatum (Cav.) Jermy — Map 107.

Tectaria caudata Cav.

Notes. Hs added (not given in Fl. Eur.), first recorded from the European mainland by B. Molesworth Allen, Lagascalia 1: 83—85 (1971), Brit. Fern Gaz. 10: 200 (1971).
Total range. Açores, Madeira, Islas Canarias, Cabo Verde (A. C. Jermy, Brit. Fern Gaz. 9: 161 (1964)).

Diplazium allorgei Tardieu-Blot — Map 108.

Total range. According to A. C. Jermy, in C. M. Ward, Brit. Fern Gaz. 10: 122 (1970), endemic to the Açores rather than an introduction from S. America.

Diplazium sibiricum (Turcz. ex G. Kunze) Kurata — Map 109.

Athyrium crenatum (Sommerf.) Rupr.

Nomenclature. Nameqata & Kurata, Enum. Jap. Pterid., p. 340 (1961).
Notes. Rs(E) added (not given in Fl. Eur.).

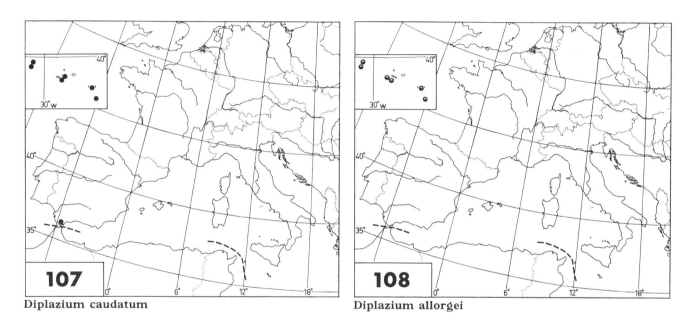

Diplazium caudatum

Diplazium allorgei

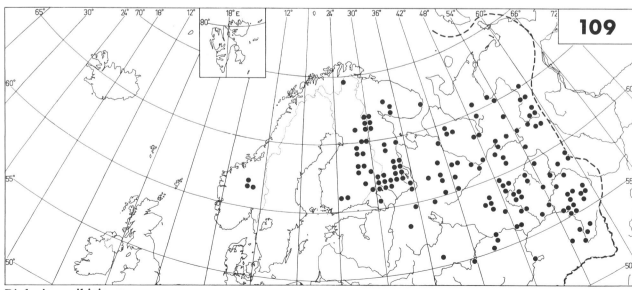

Diplazium sibiricum

Cystopteris fragilis (L.) Bernh — Map 110

Cystopteris alpina (Roth) Desv.; C. regia (L.) Desv.

Taxonomy. In addition to *C. fragilis* in the sense of Fl. Eur. (polymorphic in itself), the data from several countries include *C. dickieana*.

Notes. Bl omitted (given in Fl. Eur.), Tu added (not given in Fl. Eur.).

Total range. Hultén CP 1964: map 55.

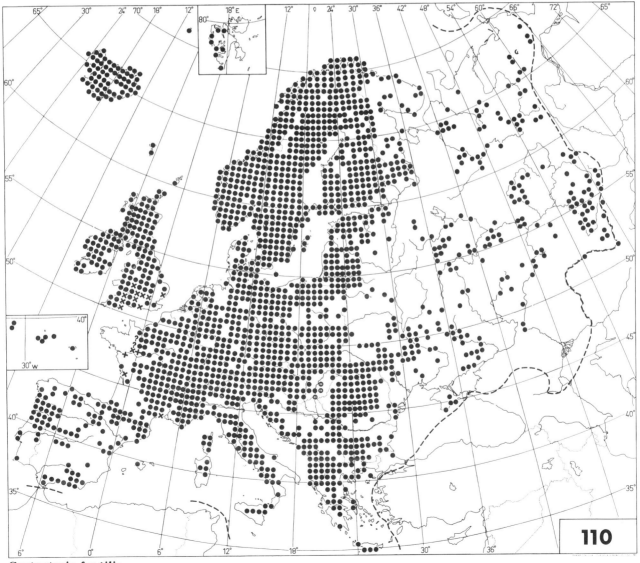

Cystopteris fragilis

Cystopteris dickieana R. Sim — Map 111.

Incl. Cystopteris baenitzii Dörfler

Taxonomy. Cf. *C. fragilis;* map provisional.
Notes. It and Sa added (not given in Fl. Eur.), Sb omitted (given in Fl. Eur.).
Total range. Hultén CP 1964: map 56.

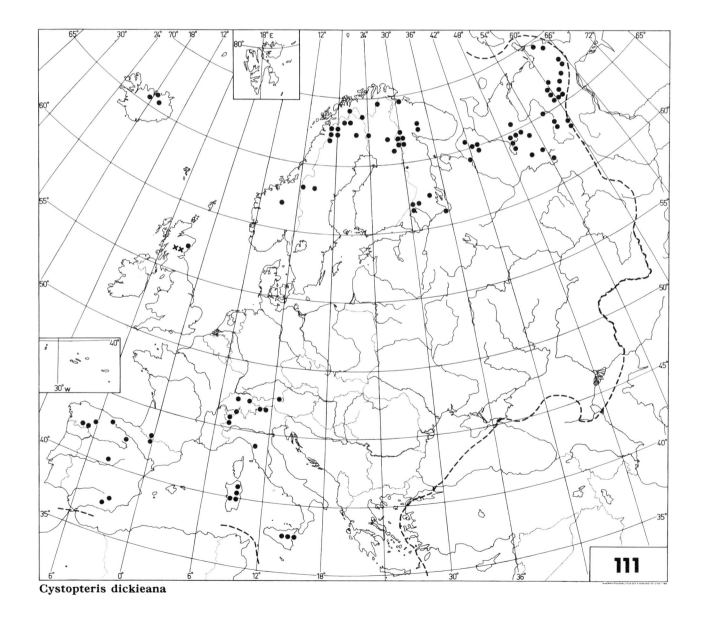

Cystopteris dickieana

Cystopteris montana (Lam.) Desv. — Map 112.

Notes. Be and Da omitted (given in Fl. Eur.), records from northwestern and central Hs omitted (given in Hultén AA 1958 and MJW 1965).

Total range. Hultén AA 1958: map 226; MJW 1965: map 15c.

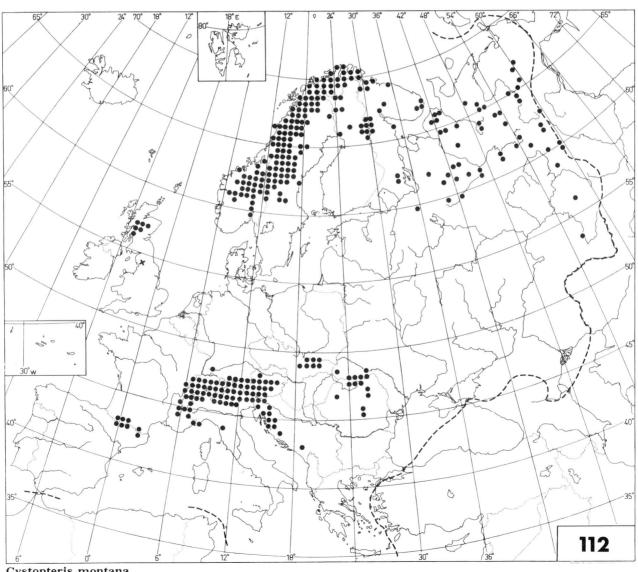

Cystopteris montana

Cystopteris sudetica A. Braun & Milde — Map 113.

Notes. †Au and Rs(B) added (not given in Fl. Eur.), Rs(C) confirmed (?Rs(C) in Fl. Eur.).

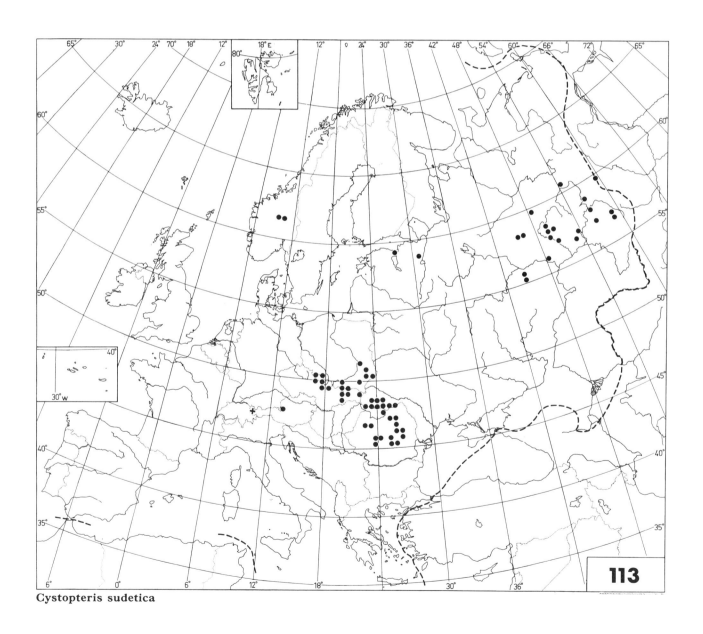

Cystopteris sudetica

Woodsia ilvensis (L.) R. Br. — Map 114.

Woodsia ilvensis subsp. rufidula (Michx) Ascherson

Notes. Ga confirmed (?Ga in Fl. Eur.), Rs(K) omitted (given in Fl. Eur.).
Total range. Hultén CP 1964: map 49; MJW 1965: map 15d.

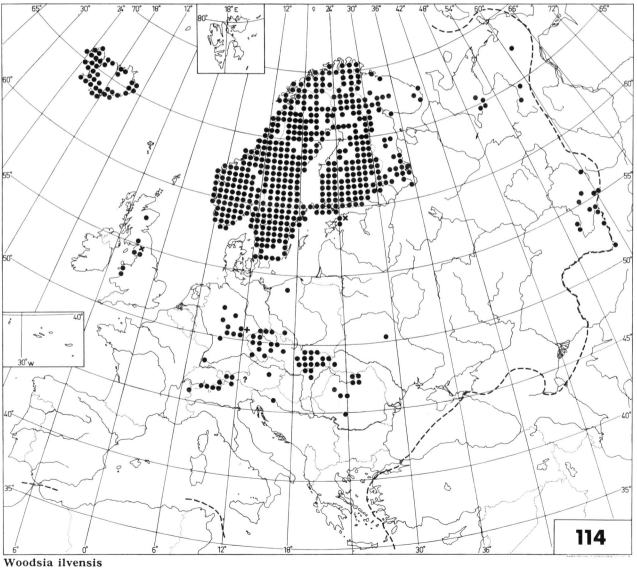

Woodsia ilvensis

Woodsia alpina (Bolton) S. F. Gray — Map 115.

Woodsia hyperborea (Liljeblad) R. Br.; W. ilvensis subsp. alpina (Bolton) Ascherson

Notes. Hu added (not given in Fl. Eur.).
Total range. Hultén AA 1958: map 210; MJW 1965: map 16a.

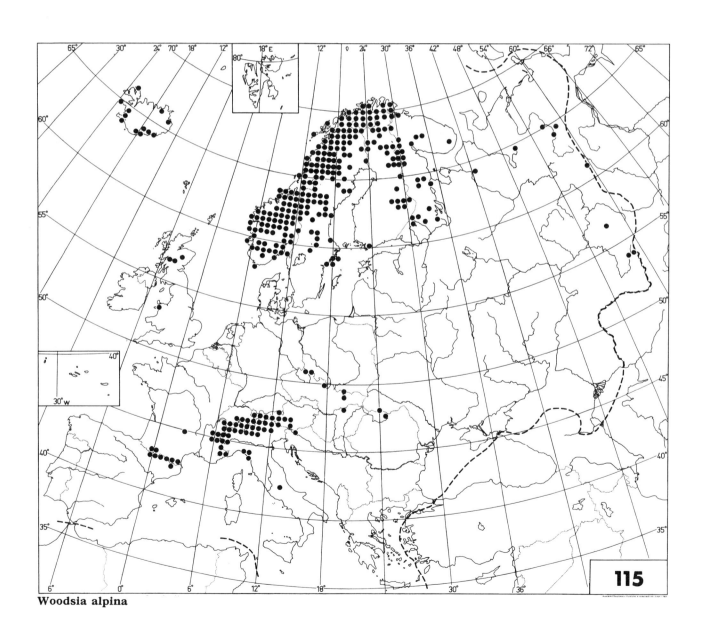

Woodsia alpina

Woodsia pulchella Bertol.   Map 116.

Woodsia glabella auct. eur. centr., non R. Br.

Taxonomy. Regarded also as conspecific with *W. glabella* (e. g. D. E. Meyer, Willdenowia 2: 214—217 (1959)), at most deserving varietal status under this (D. F. M. Brown, Beih. Nova Hedwigia 16: 79 (1964)). Reported from the Pyrenees by S. Rivas-Martínez & M. Costa, Anal. Inst. Bot. Cavanilles 26: 39—44 (1968).

Notes. Hs added (not given in Fl. Eur.).

Total range. Endemic to Europe.

Woodsia glabella R. Br. — Map 116.

Notes. Is omitted (given in Fl. Eur.), Rs(C) added (not given in Fl. Eur.).

Total range. Hultén CP 1964: map 38.

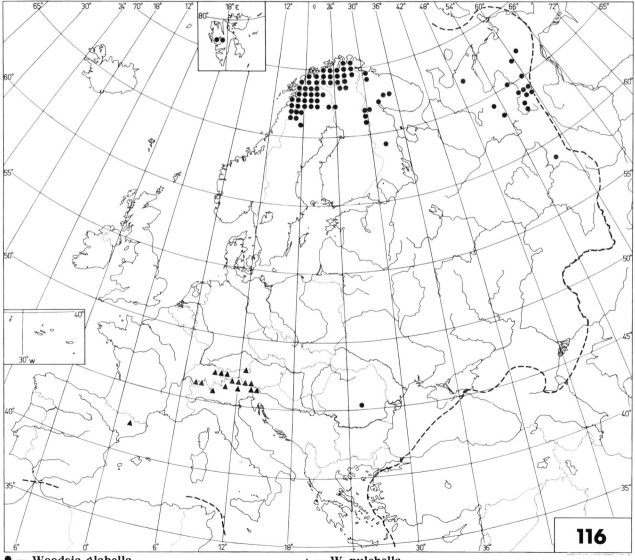

● = Woodsia glabella   ▲ = W. pulchella

Matteuccia struthiopteris (L.) Tod. — Map 117.

Struthiopteris filicastrum All.; S. germanica Willd.

Total range. Hultén CP 1964: map 115.

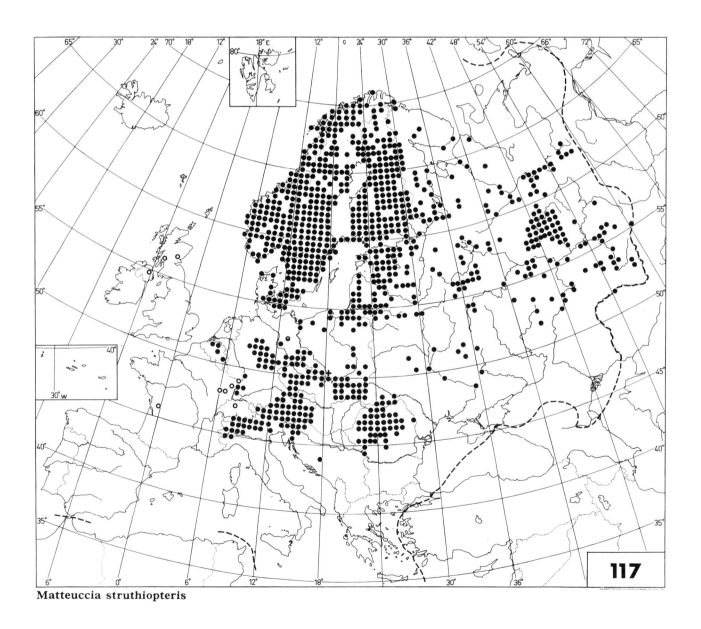

Matteuccia struthiopteris

Onoclea sensibilis L. — Map 118.

Notes. [Be] and [Ho] omitted (given in Fl. Eur.).
Native of North America and east Asia.

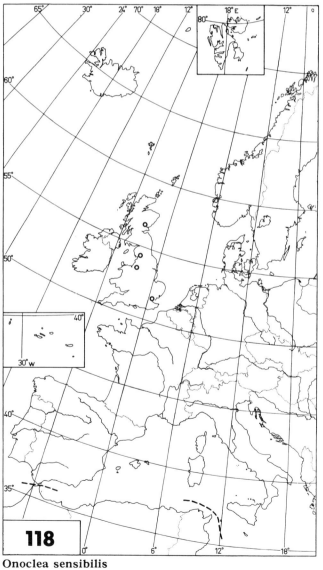

Onoclea sensibilis

ASPIDIACEAE

Polystichum lonchitis (L.) Roth — Map 119.

Aspidium lonchitis (L.) Swartz

Total range. Hultén AA 1958: map 219; MJW 1965: map 18c.

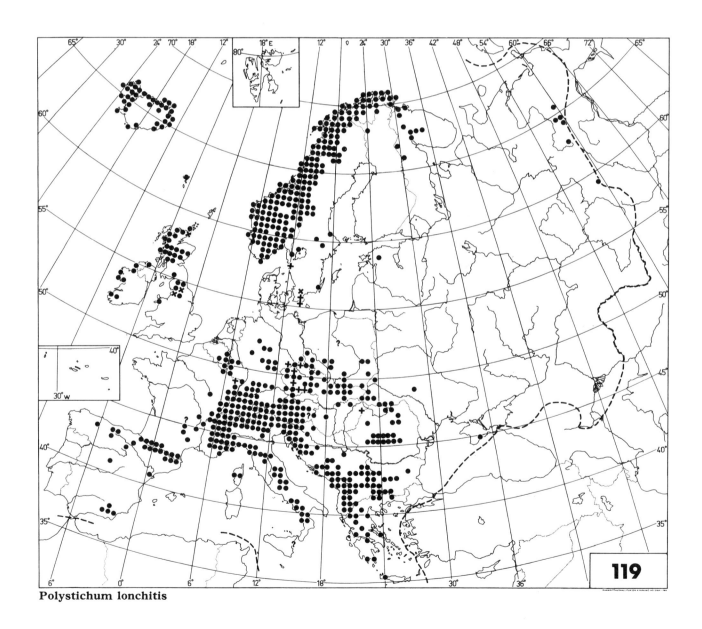

Polystichum lonchitis

Polystichum aculeatum (L.) Roth — Map 120

Aspidium lobatum (Hudson) Swartz; Polystichum lobatum (Hudson) Chevall.

Notes. In Da extinct (given as present in Fl. Eur.), †Fe added (no record in Fl. Eur.), Sa and ?Si omitted (given in Fl. Eur.), in Tu doubtful (given as certain in Fl. Eur.).

Total range. Hultén CP 1964: map 141 (including *P. setiferum*).

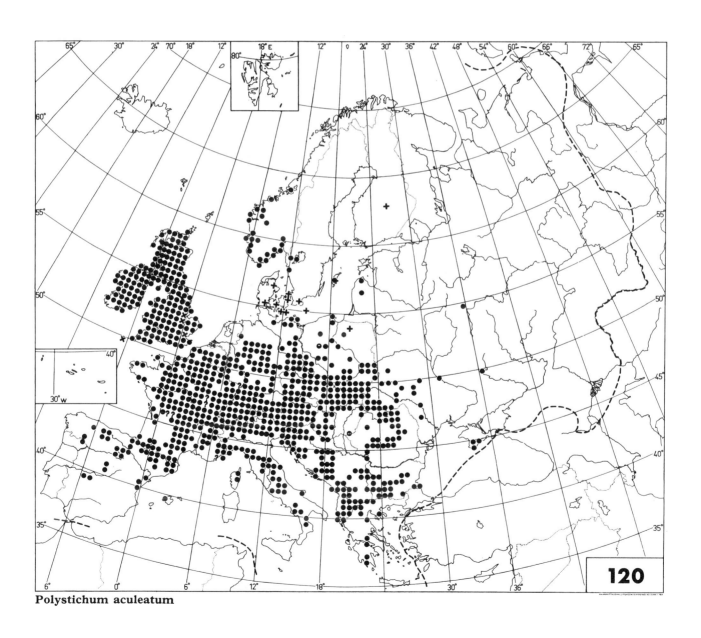

Polystichum aculeatum

Polystichum setiferum (Forsskål) Woynar — Map 121.

Aspidium aculeatum Swartz pro parte; Polystichum aculeatum auct., non (L.) Roth; P. angulare (Kit. ex Willd.) C. Presl

Notes. Cr added (not given in Fl. Eur.), Ho and ?Po omitted (given in Fl. Eur.).
Total range. R. E. G. Pichi-Sermolli, Lav. Soc. Ital. Biogeogr. 1: 106 (1971).

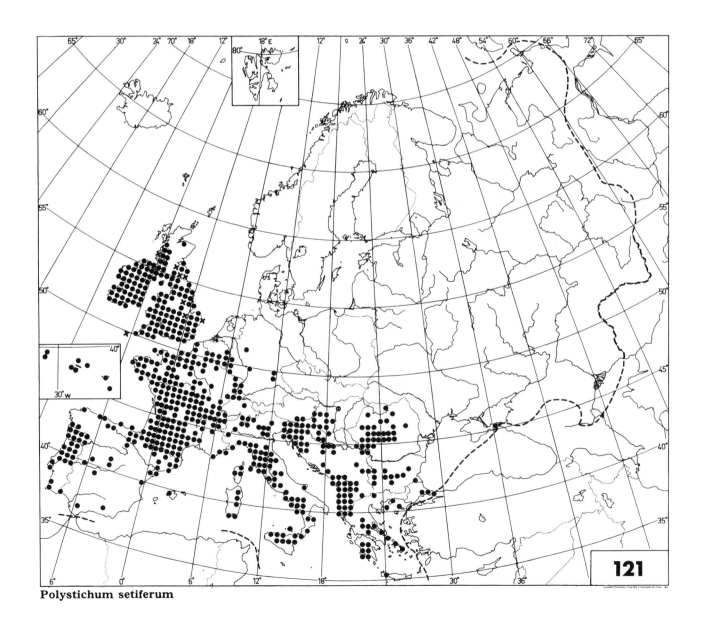

Polystichum setiferum

Polystichum braunii (Spenner) Fée — Map 122.

Aspidium braunii Spenner

Notes. Hs and Rs(E, K) added (not given in Fl. Eur.).
Total range. Hultén CP 1964: map 180.

Polystichum falcatum (L. fil.) Diels

Cyrtomium falcatum (L. fil.) C. Presl

Notes. Not mapped, although given as naturalized in Az (C. M. Ward, Brit. Fern Gaz. 10: 123 (1970)) and Br.
Native of E. Asia.

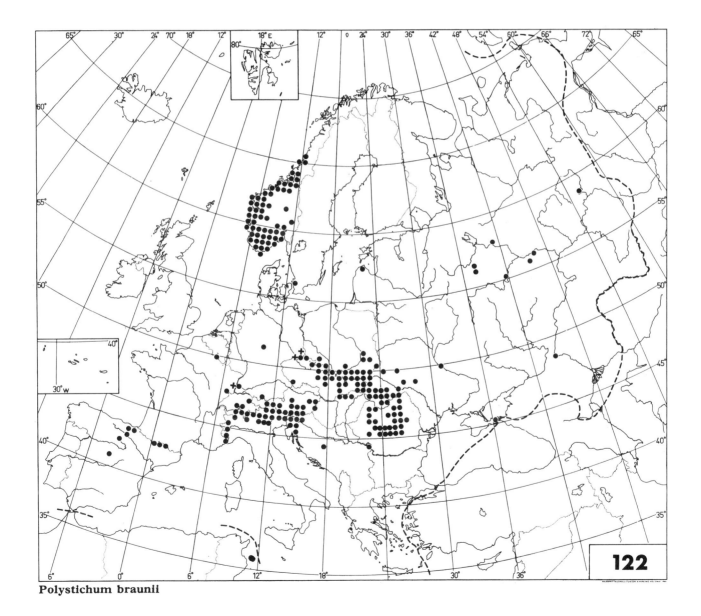

Polystichum braunii

Dryopteris filix-mas (L.) Schott — Map 123.

Nephrodium filix-mas (L.) Strempel

Notes. In Cr doubtful (given as certain in Fl. Eur.), Si added (not given in Fl. Eur.). A minor part of the records from the British Isles refer to *D. pseudomas*. However, *D. filix-mas* is probably almost ubiquitous there.
Total range. Hultén CP 1964: map 110; MJW 1965: map 17b.

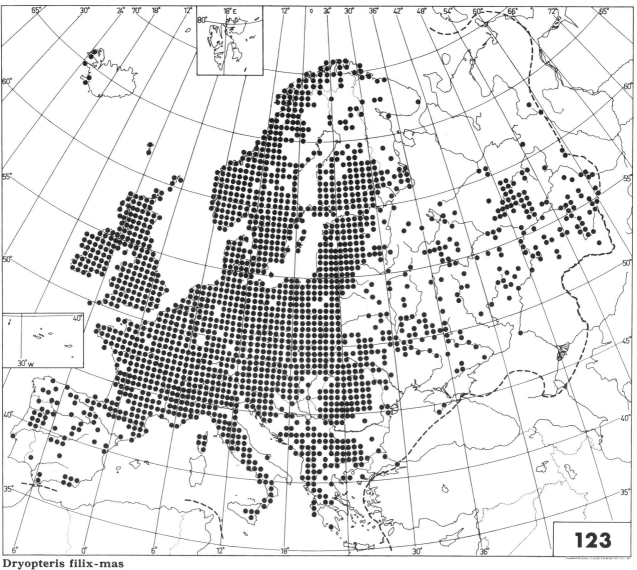

Dryopteris filix-mas

Dryopteris pseudomas (Wollaston) J. Holub & Pouzar — Map 124.

Dryopteris borreri Newman; D. paleacea (D. Don) Hand.-Mazz. pro parte, non (Swartz) C. Chr.; D. filix-mas auct. pro parte

Nomenclature. J. Holub, Folia Geobot. Phytotax. 2: 332 (1967).

Taxonomy. Part of an apogamous complex of which diploid or triploid taxa have been found in several continents. The records certainly in part refer to the apogamous, tetraploid to pentaploid *D.* × *tavelii* Rothm. (*D. filix-mas* × *D. pseudomas*).

Notes. Co, Cz and Si added (not given in Fl. Eur.), Fa omitted (?Fa in Fl. Eur.), Rm confirmed (?Rm in Fl. Eur.).

Total range. Hultén CP 1964: map 110; MJW 1965: map 17c.

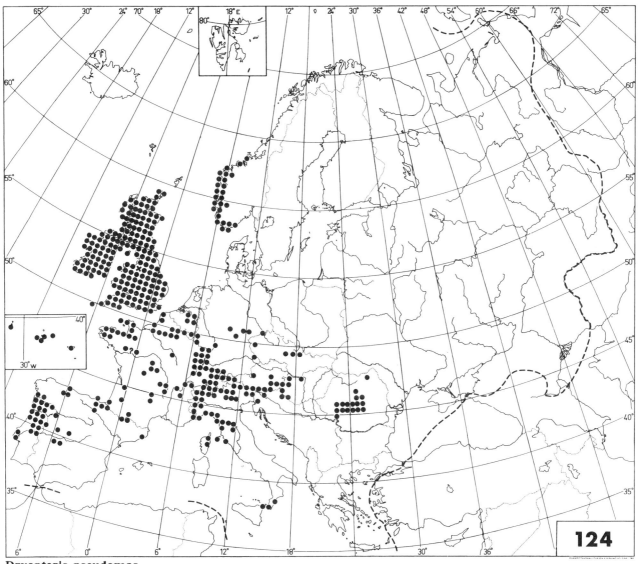

Dryopteris pseudomas

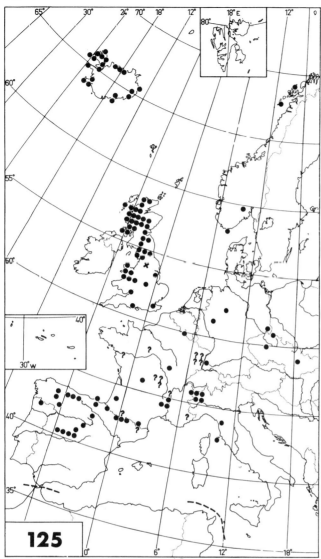

125

Dryopteris abbreviata

Dryopteris remota (A. Braun) Druce

Taxonomy. An apogamous triploid producing viable spores. From Britain, however, a sterile tetraploid hybrid (one single plant) has also been reported (I. Manton, Problems Cytol. and Evol. Pteridophyta, p. 72—73 (Cambridge 1950)). Of hybrid origin, most probably *D. assimilis* × *D. pseudomas;* cf. W. Gätzi, Ber. St. Gall. Naturwiss. Ges. 77: 48—61 (1961), T. Reichstein, Brit. Fern Gaz. 9: 232—233 (1965), C.-J. Widén et al., Helv. Chim. Acta 53: 2176—2177 (1970), 54: 2842—2843 (1971).

Notes. Widespread in central Europe but rather rare. Not included in Fl. Eur., and no map can be given here. May have been included, if anywhere, in the records of *D. pseudomas.*

Total range. As far as is known, endemic to Europe. A distribution map is given by W. Döpp, Planta 20: 522 (1030).

Dryopteris abbreviata (DC.) Newman — Map 125.

Notes. Cz and Ge added (not given in Fl. Eur.), Fa and Lu omitted (?Fa and ?Lu in Fl. Eur.). The map is certainly incomplete as to several countries.

Dryopteris villarii (Bellardi) Woynar ex Schinz & Thell.

Dryopteris rigida (Swartz) A. Gray; Nephrodium villarii (Bellardi) G. Beck; Polystichum rigidum (Swartz) DC.

Taxonomy. Both subsp. *villarii* and subsp. *pallida* are diploid and not always easily distinguishable from each other. In addition to the British tetraploid (G. Panigrahi, Amer. Fern Jour. 55: 1—8 (1965); O. L. Gilbert, Brit. Fern Gaz. 9: 263—268 (1966)), a tetraploid has recently been found in Rm (G. Vida, Bot. Közlem. 56: 11—15 (1969)). C.-J. Widén et al., Helv. Chim. Acta 54: 2824—2850 (1971), also report a tetraploid from Ju and found the British and Balkan tetraploids chemically similar, both probably being allotetraploids and derived from hybridization between subspp. *villarii* and *pallida.*

Notes. Rm added (not given in Fl. Eur.).

D. villarii subsp. **villarii** — Map 126.

D. villarii subsp. **pallida** (Bory) Heywood — Map 127.

Notes. Local variants occur in Bl and Hs (Sierra Nevada). The records from Al and It possibly include subsp. *villarii.*

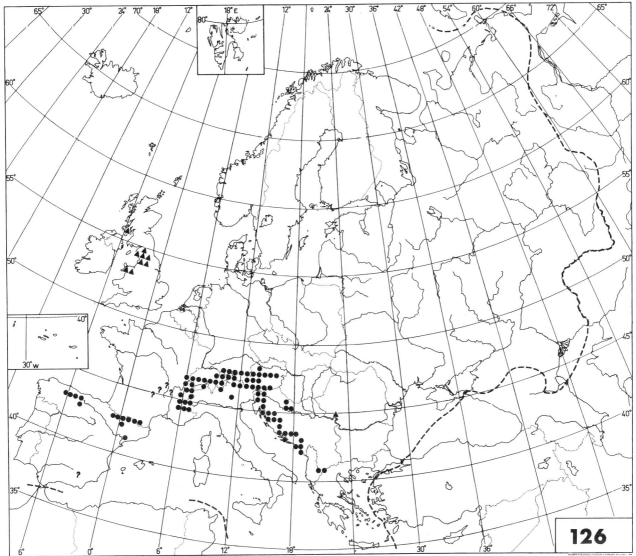

126

● = Dryopteris villarii subsp. **villarii**   ▲ = tetraploid **D. villarii**

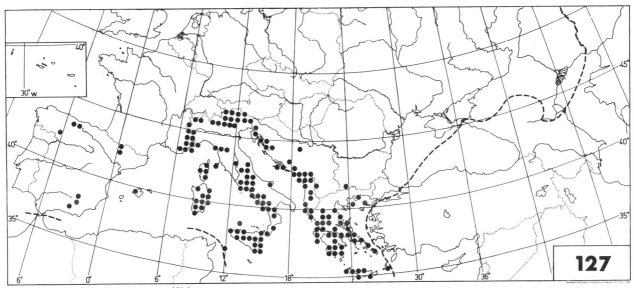

127

Dryopteris villarii subsp. pallida

Dryopteris cristata (L.) A. Gray — Map 128.

Notes. Hs added (not given in Fl. Eur.).
Total range. Hultén AA 1958: map 40; MJW 1965: map 17d.

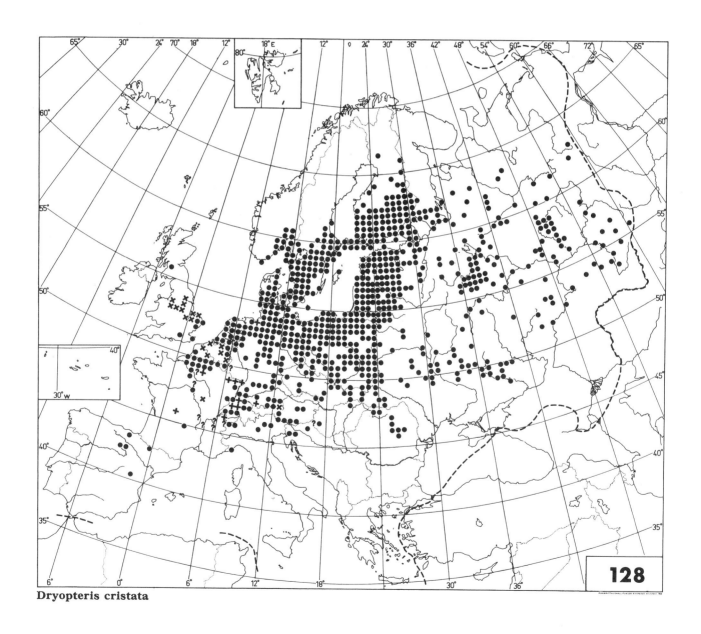

Dryopteris cristata

128

Dryopteris carthusiana (Vill.) H. P. Fuchs — Map 129.

Dryopteris spinulosa Watt.

Notes. Hs confirmed (?Hs in Fl. Eur.).
Total range. Hultén AA 1958: map 155; MJW 1965: map 18b.

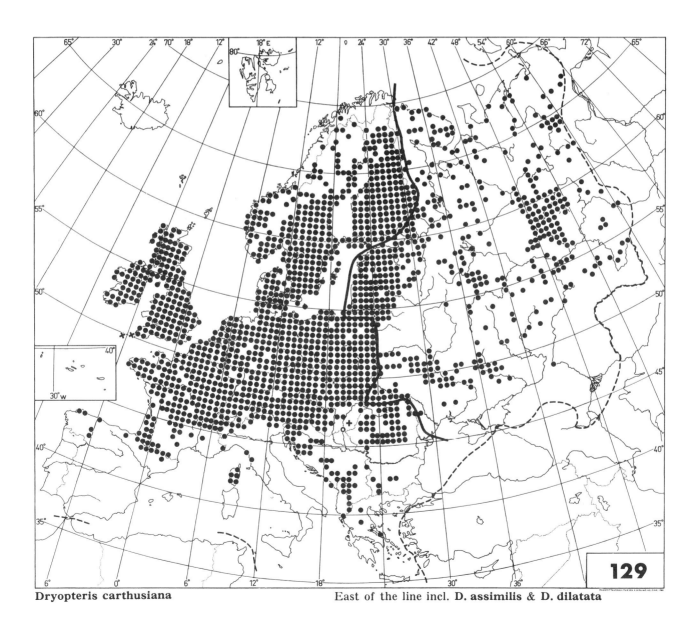

Dryopteris carthusiana   East of the line incl. **D. assimilis & D. dilatata**

129

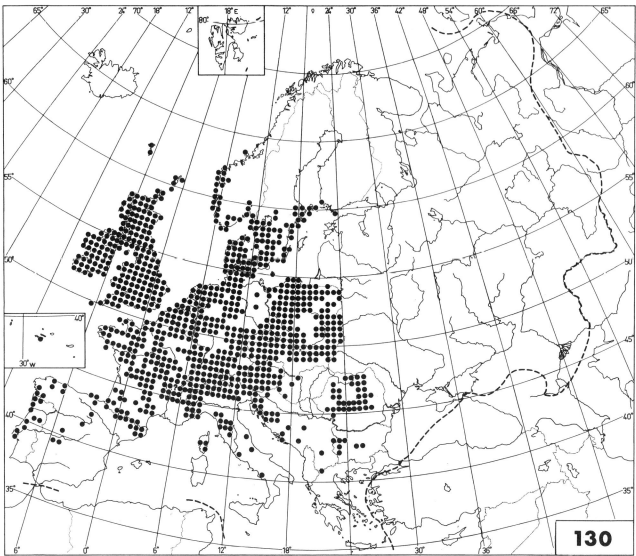

Dryopteris dilatata

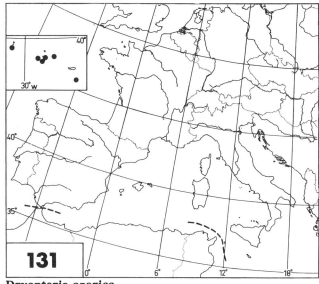

131

Dryopteris azorica

Dryopteris dilatata (Hoffm.) A. Gray — Map 130.

Dryopteris austriaca auct.

Notes. Is omitted (given in Fl. Eur.).
Total range. Hultén AA 1958: map 156; MJW 1965: map 18a (both including *D. assimilis*). Outside Europe possibly only in Asia Minor and the Caucasus.

Dryopteris azorica (Christ) Alston — Map 131.

Total range. Endemic to the Açores.

Dryopteris assimilis S. Walker — Map 132.

Taxonomy. Not until 1961 specifically separated from *D. dilatata*. In many parts of their overlapping areas the records are still inadequate and uneven. Some of the records given by T. Simon & G. Vida, Ann. Univ. Sci. Budapest, Sect. Biol. 8: 275—284 (1966), were not included in the material provided by certain countries. J. A. Crabbe et al., Watsonia 8: 3—15 (1970), dealt especially with spore morpohology and the British distribution, whilst C.-J. Widén et al., Acta Bot. Fenn. 91: 1—30 (1970), studied the chemotaxonomical pattern of *D. assimilis* and several related species.

Notes. Au, ?Bu, Cz, Da, Fe, Is, Ju and Rm added (not given in Fl. Eur.).

Total range. Given as endemic to Europe in Fl. Eur. However, plants identical with the European ones occur in western North America and eastern Asia, and nearby taxa are found in eastern North America; C.-J. Widén & D. M. Britton, Canad. Jour. Bot. 49: 247—258 (1971).

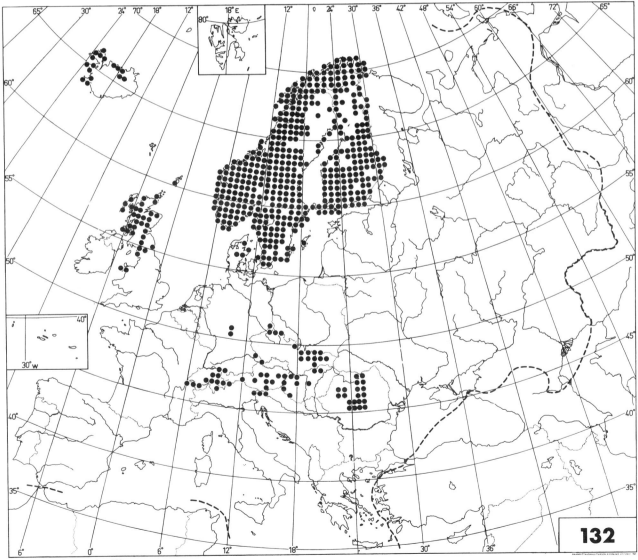

Dryopteris assimilis

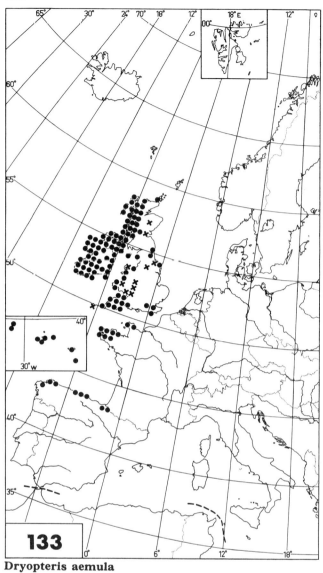

133

Dryopteris aemula

Dryopteris aemula (Aiton) O. Kuntze — Map 133.

Total range. Europe, Açores, Madeira, Islas Canarias (A. C. Jermy, Brit. Fern Gaz. 10: 9—12 (1968)).

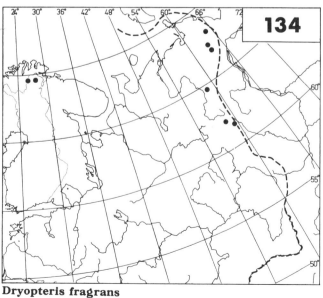

134

Dryopteris fragrans

Dryopteris fragrans (L.) Schott — Map 134.

Total range. Hultén CP 1964: map 27.

Gymnocarpium dryopteris (L.) Newman — Map 135.

Carpogymnia dryopteris (L.) Á. & D. Löve; Dryopteris linneana C. Chr.; Lastrea dryopteris (L.) Bory; Nephrodium dryopteris (L.) Michx; Phegopteris dryopteris (L.) Fée; Polypodium dryopteris L.

Nomenclature. Á. & D. Löve, Taxon 16: 191—192 (1967); R. E. Holttum, Taxon 17: 529—530 (1968); C. V. Morton, Taxon 18: 661—662 (1969).

Notes. Cr and Rs(K) omitted (given in Fl. Eur.).

Total range. Hultén CP 1964: map 108; MJW 1965: map 16c.

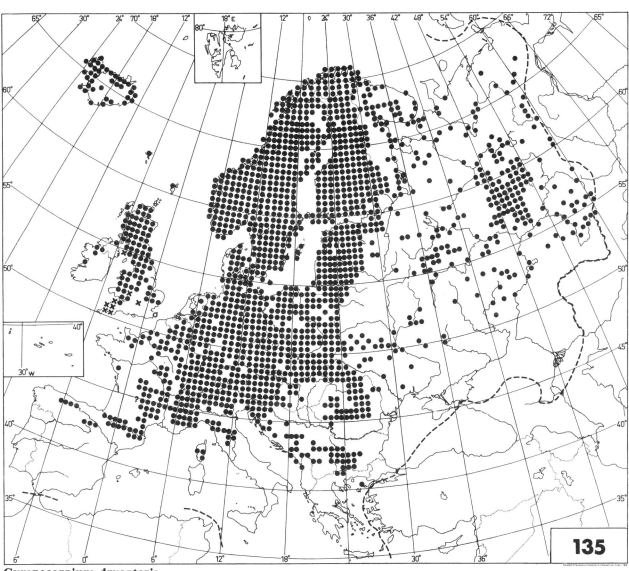

Gymnocarpium dryopteris

Gymnocarpium robertianum (Hoffm.) Newman — Map 136.

Carpogymnia robertiana (Hoffm.) Á. & D. Löve; Dryopteris robertiana (Hoffm.) C. Chr.; Lastrea robertiana (Hoffm.) Newman; Nephrodium robertianum (Hoffm.) Prantl; Phegopteris robertiana (Hoffm.) A. Braun; Polypodium robertianum Hoffm.

Nomenclature. Cf. *Gymnocarpium dryopteris.*

Taxonomy. Includes the plants from NE. Europe provisionally identified as *Gymnocarpium continentale* (Petrov) Pojark. or *G. heterosporum* Wagner; cf. A. I. Tolmachev, Arkt. Fl. SSSR. 1: 29 (1960), and P. Kallio et al., Ann. Univ. Turkuensis A II 42 (Rep. Kevo Subarctic Res. Stat. 5): 96—97 (1969).

Notes. In Co extinct (given as present in Fl. Eur.).

Total range. Hultén CP 1964: map 116; MJW 1965: map 16b.

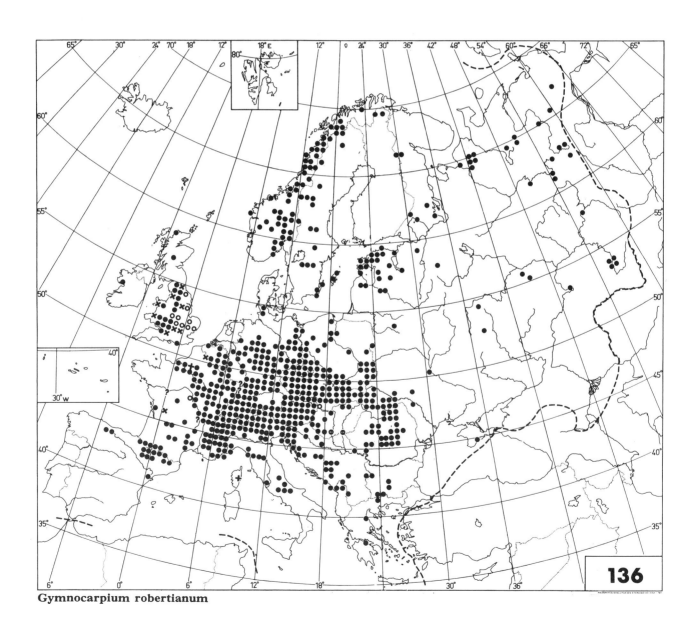

Gymnocarpium robertianum

ELAPHOGLOSSACEAE

Elaphoglossum paleaceum (Hooker & Grev.)
Sledge — Map 137.

Acrostichon paleaceum Hooker & Grev.; Elaphoglossum hirtum auct.

Nomenclature. W. Sledge, Bull. Brit. Mus. (Nat. Hist.) 4(2): 95 (1967); R. E. G. Pichi-Sermolli & E. A. Schelpe, Webbia 23: 149—151 (1968).

Total range. Açores and Madeira.

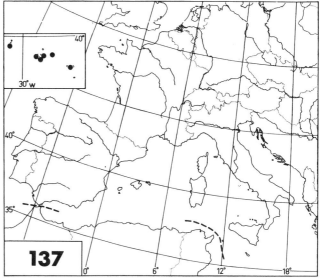

Elaphoglossum paleaceum

BLECHNACEAE

Woodwardia radicans (L.) Sm. — Map 138.

Notes. Co and Cr added (not given in Fl. Eur.).
Total range. R. A. Fataliyev, Bot. Žur. 46: 1317 (1961).

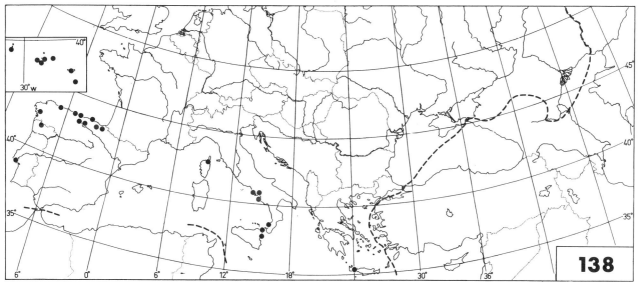

Woodwardia radicans

Blechnum spicant (L.) Roth — Map 139.

Taxonomy. For infraspecific variation in Is, Hs and Lu, see Á. & D. Löve, Bot. Tidsskr. 62: 186—196 (1966), Collect. Bot. (Barcelona) 7: 665—676 (1968).

Notes. Al added (not given in Fl. Eur.), Bl omitted (given in Hultén CP 1964).

Total range. Hultén CP 1964: map 143; MJW 1965: map 12c; R. E. G. Pichi-Sermolli, Lav. Soc. Ital. Biogeogr. 1: 114 (1971).

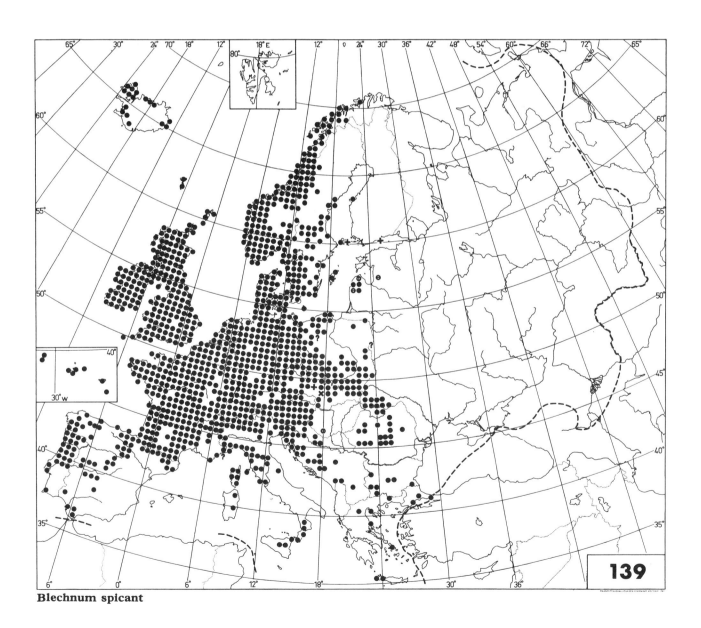

Blechnum spicant

POLYPODIACEAE

Polypodium australe Fée — Map 140.

Polypodium serratum (Willd.) Sauter, non Aublet; P. vulgare subsp. serrulatum Arcangeli

Taxonomy. Cf. *P. vulgare*. Certainly under-represented in many areas in the map.
Notes. Al, Az, Br, Bu, Hb and Rs(K) omitted (all given in Fl. Eur.).
Total range. Hultén CP 1964: map 167A; MJW 1965: map 18d.

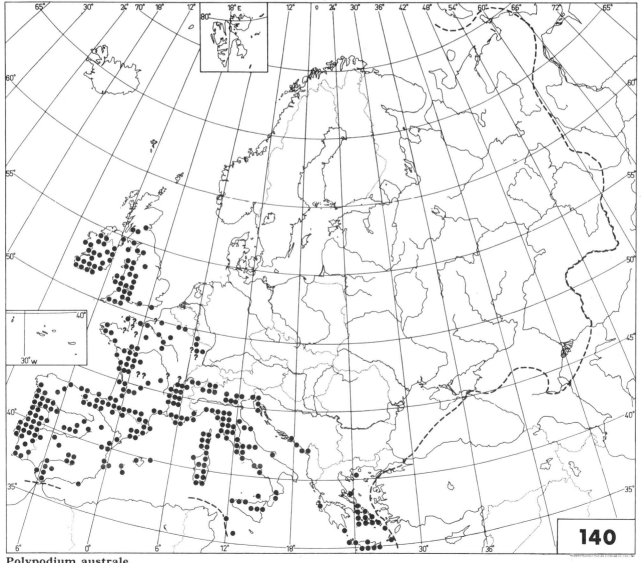

Polypodium australe

Polypodium vulgare L., sensu lato — Map 141.

Taxonomy. The tetraploid *P. vulgare*, sensu stricto, is not readily distinguishable from the diploid *P. australe* and the hexaploid *P. interjectum*, in particular as regards older literature records. Interspecific hybrids are not uncommon. Records for *P. vulgare* from several countries include *P. vulgare* coll.; consequently, no map of *P. vulgare*, sensu stricto, can be given here.

Notes. In addition to the countries given in Fl. Eur., *P. vulgare* s. str. is recorded for Tu, whereas the records for Az refer to *P. azoricum*.

Total range. Hultén CP 1964: map 167A; MJW 1965: map 18d (both including *P. interjectum*, in particular).

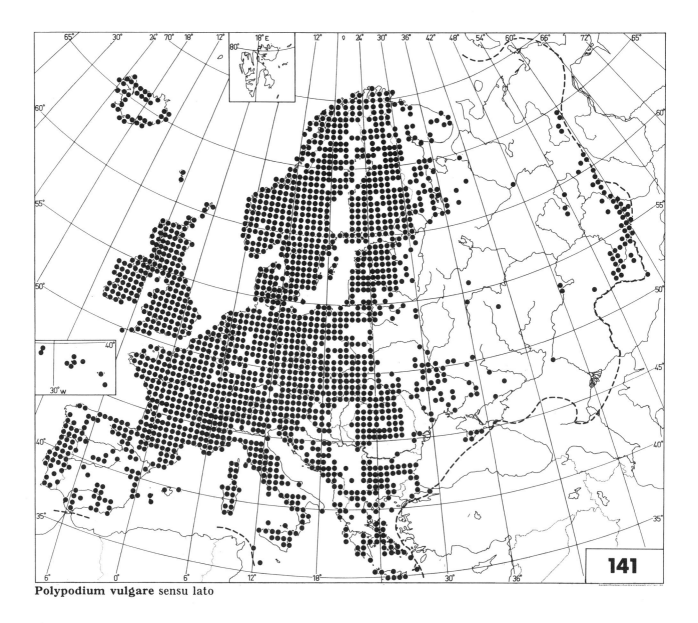

Polypodium vulgare sensu lato

Polypodium interjectum Shivas — Map 142.

Polypodium vulgare subsp. prionodes (Ascherson) Rothm.

Taxonomy. Cf. *P. vulgare.* Certainly under-represented in many areas in the map.
Notes. Au, Co, Cz and Rs(B, W, K) added (not given in Fl. Eur.).

Polypodium azoricum (Vasc.) Fernandes — Map 142.

Taxonomy and nomenclature. R. Fernandes, Bol. Soc. Brot. 42: 241—247 (1968), refers the *Polypodium* material from the Açores to this species.

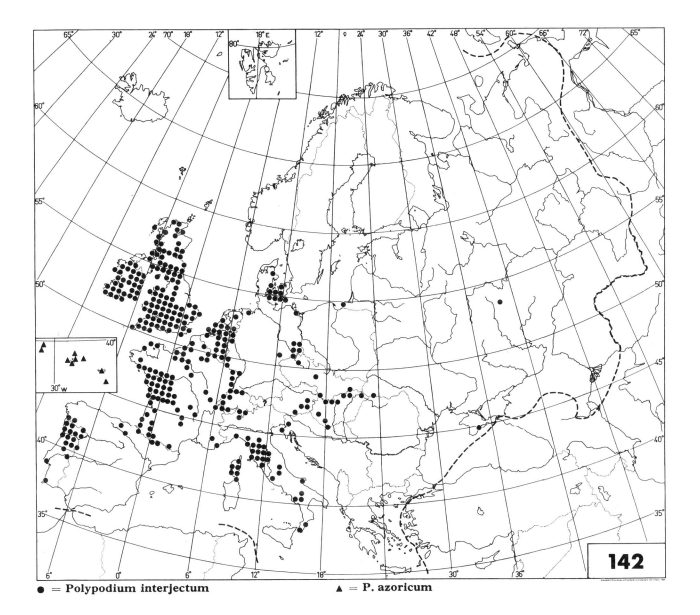

● = Polypodium interjectum   ▲ = P. azoricum

MARSILEACEAE

Marsilea quadrifolia L. — Map 143.

Notes. Az added (not given in Fl. Eur.), in Po extinct (given as present in Fl. Eur.).

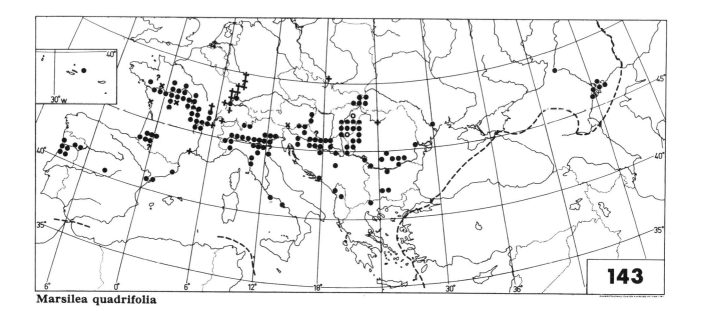

Marsilea quadrifolia

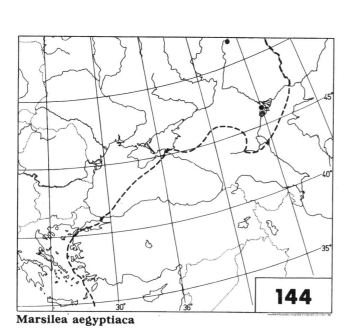

Marsilea aegyptiaca

Marsilea aegyptiaca Willd. — Map 144.

Notes. Rs(W) omitted (given in Fl. Eur.).

Marsilea strigosa Willd. — Map 145.

Marsilea pubescens Ten.

Notes. Hs confirmed (?Hs in Fl. Eur.).

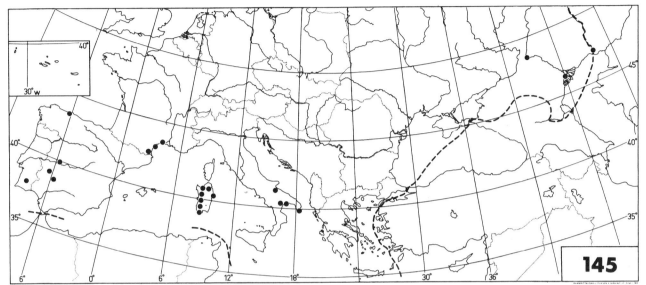

Marsilea strigosa

Pilularia minuta Durieu ex A. Braun — Map 146.

Notes. Co added (not given in Fl. Eur.), Hs omitted (?Hs in Fl. Eur.).

Total range. MJW 1965: map 19a.

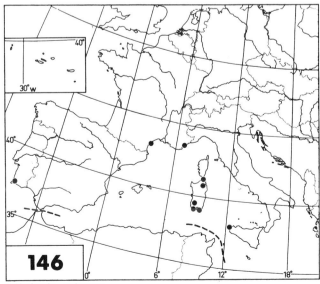

Pilularia minuta

Pilularia globulifera L. — Map 147.

Notes. In He found alive in 1970 (†He in Fl. Eur.).
Total range. Endemic to Europe.

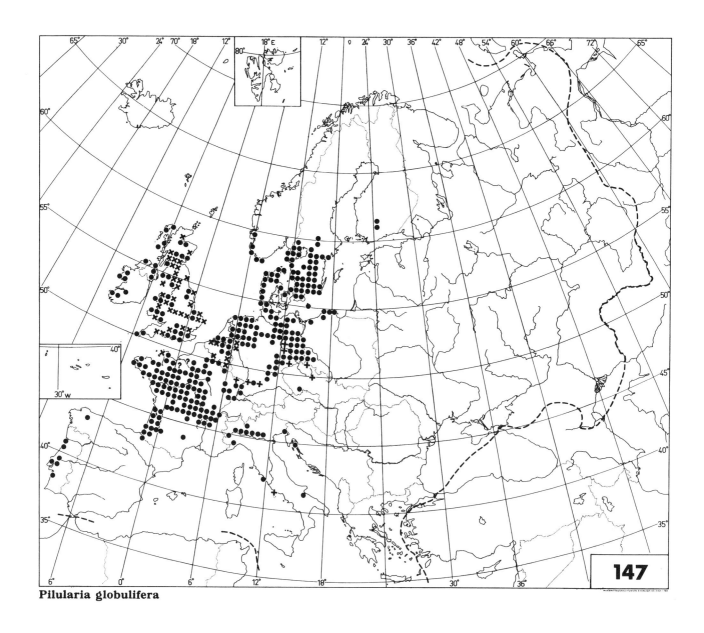

Pilularia globulifera

147

SALVINIACEAE

Salvinia natans (L.) All. — Map 148.

Notes. In Ho extinct (given as present in Fl. Eur.), Rs(B, C) and Tu added (not given in Fl. Eur.), Rs(K) and Si omitted (given in Fl. Eur.).

Total range. K. K. Shaparenko, Paleobotanika 2, Bot. Inst. V. L. Komarova Akad. Nauk SSSR., ser. 8: 34 (1956); MJW 1965: map 19b.

Salvinia rotundifolia Willd.

Notes. [Hs] omitted (given in Fl. Eur.). Consequently, this central and South American plant is nowhere established in Europe.

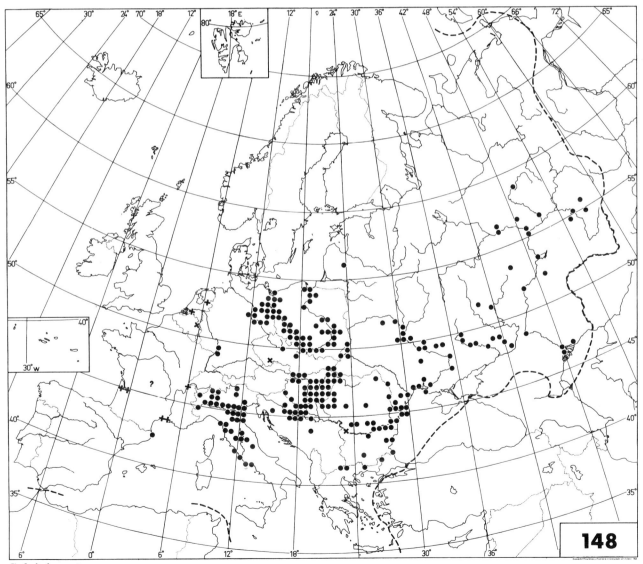

Salvinia natans

AZOLLACEAE

Azolla filiculoides Lam. — Map 149.

Notes. [Hs] omitted (given in Fl. Eur.), [Ju] added (not given in Fl. Eur.).

Total range. Native of tropical America. R. Herzog, Bot. Arch. (Königsberg) 39: 224 (1938); H. Tralau, Bot. Not. 112: 387—389 (1959).

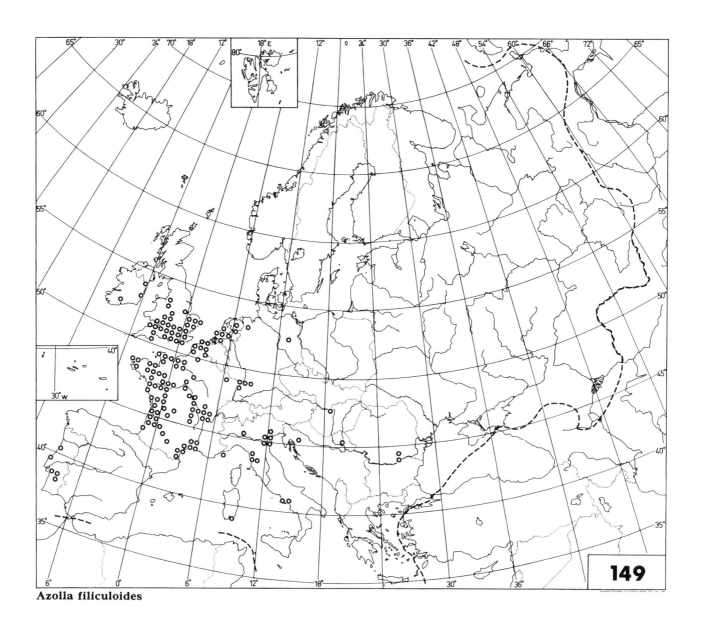

Azolla filiculoides

Azolla caroliniana Willd.   Map 150.

Notes. [Be] and [Hs] omitted (given in Fl. Eur.).
Total range. Native of America. R. Herzog, Bot. Arch. (Königsberg) 39: 224 (1938).

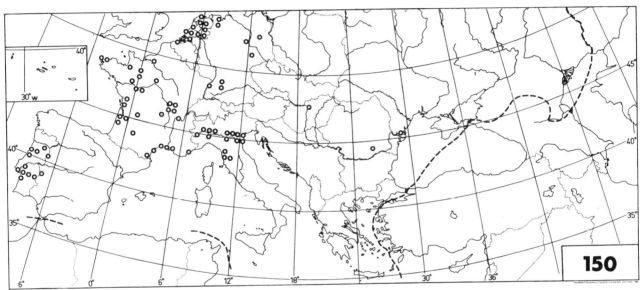

Azolla caroliniana

2
GYMNOSPERMAE
(PINACEAE TO EPHEDRACEAE)

CONTENTS

THE COMMITTEE FOR MAPPING THE FLORA OF EUROPE

SECRETARIAT, HELSINKI

J. JALAS (Chairman of the Committee)
J. SUOMINEN (Secretary General)
U. KURIMO (Technical Assistant)

ADVISERS

T. W. BÖCHER, København
A. R. CLAPHAM, Sheffield
E. HULTÉN, Stockholm
D. A. WEBB, Dublin

COMMITTEE MEMBERS ACTING AS REGIONAL COLLABORATORS

Albania (Al)	I. MITRUSHI, Tiranë	
Austria (Au)	F. EHRENDORFER, Wien	H. NIKLFELD, Wien
Belgium (Be, excl. Luxembourg)	E. VAN ROMPAEY, Antwerpen	
British Isles (Br, Hb, Channel Islands)	F. H. PERRING, Abbots Ripton	
Bulgaria (Bu)	S. KOŽUHAROV, Sofia	
Czechoslovakia (Cz)	J. FUTÁK, Bratislava	J. HOLUB, Průhonice
Denmark (Da, Fa)	A. HANSEN, København	
Finland (Fe)	J. SUOMINEN, Helsinki	
France (Co, Ga)	P. DUPONT, Nantes Maps checked by several French botanists	

German Democratic Republic (Ge: DDR)	H. Meusel, Halle Assisted by: W. Fischer, Potsdam F. Fukarek, Greifswald W. Hempel, Dresden	E. Weinert, Halle H.-D. Krausch, Potsdam W. Müller-Stoll, Potsdam J. Pötsch, Potsdam
German Federal Republic (Ge: BRD)	H. Ellenberg, Göttingen P. Schönfelder, Stuttgart Assisted by collaborators in the mapping of the Central European Flora	H. Haeupler, Göttingen
Greece (Cr, Gr)	P. Critopoulos, Athínai Assisted by: S. Dafis, Thessaloniki W. Greuter, Genève (Kriti)	G. Lavrentiades, Thessaloniki H. Runemark, Lund (Aegean Islands)
Hungary (Hu)	Z. E. Kárpáti (†), Budapest Assisted by: A. Terpó, Budapest	R. v. Soó, Budapest
Iceland (Is)	E. Einarsson, Reykjavík	
Italy (It, Sa, Si)	E. Nardi, Firenze	G. Moggi, Firenze
Jugoslavia (Ju)	E. Mayer, Ljubljana	
Luxembourg (Be, excl. Belgium)	L. Reichling, Luxembourg	
Netherlands (Ho)	J. Mennema, Leiden	
Norway (No, Sb)	K. Faegri, Bergen	
Poland (Po)	M. Gostyńska, Kórnik Assisted by: D. Fijałkowski, Lublin J. Guzik, Kraków J. Jasnowska, Szczecin K. Kępczyński, Toruń M. Kopij, Warszawa T. Krzaczek, Lublin J. Madalski, Wrocław R. Olaczek, Łódź	J. Kornaś, Kraków L. Olesiński, Olsztyn Z. Schwarz, Gdańsk A. Sokołowski, Białowieża R. Sowa, Łódź H. Rutowicz, Łódź A. Zając, Kraków E. U. Zając, Kraków W. Żukowski, Poznań
Portugal (Az, Lu)	J. do Amaral Franco, Lisboa Assisted by: M. L. da Rocha-Afonso, Lisboa	
Romania (Rm)	A. Borza (†), Cluj	N. Boşcaiu, Cluj

6

Spain (Bl, Hs)	E. F. Galiano, Sevilla Assisted by: P. Montserrat, Jaca	B. Valdés, Sevilla
Sweden (Su)	E. Hultén, Stockholm	
Switzerland (He)	O. Hegg, Bern	M. Welten, Bern
Turkey (European part; Tu)	D. A. Webb, Dublin	
U.S.S.R. (Rs (N, B, C, W, K, E))	A. I. Tolmachev, Leningrad Assisted by: I. Musaev, Leningrad T. Plieva, Leningrad	O. Svjazeva, Leningrad

FINNISH CONSULTATION COMMITTEE, HELSINKI

T. Ahti
P. Isoviita
J. Jalas (Chairman)
A. Kalela

I. Kukkonen
H. Luther
R. Ruuhijärvi
J. Suominen (Secretary)

PREFACE

After several experimental stages and diverse preparatory work performed both within the Secretariat and by individual members of the Committee for Mapping the Flora of Europe, the first volume of »Atlas Florae Europaeae», dealing with the pteridophytes (Psilotaceae to Azollaceae; 121 pp., 150 maps), finally came from the printer's on May 19, 1972.

Originally, it was planned to include the gymnosperms in this first volume. However, as this would have meant some additional delay, it was decided to publish the maps for the representatives of this fairly limited and taxonomically well-defined group as a separate volume. This arrangement may also be convenient for the users of »Atlas Florae Europaeae», as some of them may be supposed to be specialists in one or other of these two major taxonomical groups, and will not necessarily be particularly interested in the other.

Apart from some minor improvements or adjustments explained in »Explanatory notes» below, and chiefly caused by the special nature of this taxonomic group, the species maps and textual comments in the present volume have been prepared along the same lines as those of the first one. For details of the mapping procedure, pages 9 to 13 of Vol. 1 should be consulted.

With the present volume, the European mapping scheme can be considered to be well past its initial phase and the technical difficulties encountered at the start. May this be an inspiration to us to continue the fruitful cooperation of our multi-national European team.

EXPLANATORY NOTES

The species and subspecies included (natives only; see p. 8) are according to Flora Europaea (Vol. 1, 1964). However, in a few cases a slight modification of the delimitation and/or nomenclature of the taxa proved desirable, mainly because of new data published after 1964, or details in the preliminary maps received from the individual European countries. A collective species map has been included when the maps of the subspecies are not entirely satisfactory, either because of disagreement regarding the delimitation of the subspecific entities, or because of partial lack of detailed information. In other cases, new recently published views and facts deviating from the treatment in Flora Europaea are dealt with in the text under the headings *Taxonomy* and *Nomenclature.*

When appropriate, the comments on taxonomy and nomenclature are supplemented by a selection of *synonyms*, including those appearing in the text of Flora Europaea, and, in particular, combinations formed in the most recent literature or otherwise representing different views on the status and affinities of the taxon involved.

Under the heading *Notes*, important new or omitted records are indicated (mainly by giving the territorial abbreviations), especially when they complete or correct Flora Europaea. When the distributional information given by Flora Europaea is not particularly detailed, as is sometimes the case at the level of subspecies, all the pertinent territories are listed. As it is evident that the number of new or omitted records tends to increase with the age of the corresponding volume of Flora Europaea, some other solution may be better suited for the future volumes of Atlas Florae Europaeae.

Total range is treated in two different ways. In the case of European endemics, the total range is presented in the maps, and, consequently, no references are given to distribution maps published earlier. Species and subspecies endemic to areas within continental Europe are simply given as »endemic to Europe», whereas for those restricted to geographically well-separated areas, such as islands, more detailed information is provided, e.g. »endemic to the Açores». On the other hand, when the total range exceeds the European boundaries, original total range maps have been cited, if any. When the taxon has been mapped by Hultén CP 1964 or MJW 1965 (for the abbreviations, see below), references to earlier maps have generally been omitted. Of the maps published in the more recent literature, reference has preferably been made to those essentially completing, rectifying or adjusting the picture of the total range. For a more complete list of published distribution maps, Index holmensis, Vol. 1 (ed. by H. Tralau, 264 pp. Zürich 1969) should be consulted.

The following abbreviations are used for frequently quoted literature:

Fenaroli 1967 = L. Fenaroli, Gli Alberi d'Italia. — 320 pp. Milano 1967.
Fl. Eur. = T. G. Tutin, V. H. Heywood et al. (ed.), Flora Europaea 1. — xxxiv + 464 pp. Cambridge 1964.
Hultén CP 1964 = E. Hultén, The Circumpolar Plants. I. — Kungl. Svenska Vet.-Akad. Handl., Ser. 4, 8 (5): 1—280. 1964.
MJW 1965 = H. Meusel, E. Jäger & E. Weinert, Vergleichende Chorologie der zentraleuropäischen Flora. — Text 583 pp., Karten 258 pp. Jena 1965.

The mapping unit used is the 50-km square of the UTM (Universal Transverse Mercator) grid maps, covering Europe on a scale of 1: 1 000 000. Along the compensation zones of every sixth meridian, the breadth (W-E) of the squares varies between 40 and 60 km. Slight deviations from the UTM grid, in respect of coastal areas and islands far from the coast, are explained on p. 10 of Vol. 1, and will not be mentioned again here. In collecting the primary data in the various countries, and in compiling and publishing the final maps, a *base map*, 1 : 10 000 000, is used. A sample specimen of this base map was distributed as an appendix to Vol. 1. The final scale of the published maps is ca. 1 : 31 000 000. It may be noted that the Açores and Spitsbergen are included as insets.

The mapping symbols are:

● native occurrence
O introduction (established alien). Not used in this volume; see below
◒ status unknown or uncertain
✚ extinct
✖ probably extinct, or, at least, not recorded since 1930
? record uncertain as to identification or locality

Any reliable record was used when entering a symbol in a square, not only those based on herbarium specimens but also reliable published or unpublished sight records.

It has been agreed by the Committee to omit rare and/or ephemeral casuals, short-distance and/or inconstant escapes from cultivation, and even extensive plantations of forest trees. The last-mentioned omission concerns the present volume in particular, since it follows that cultivated trees (and shrubs) have been disregarded irrespective of the purpose of cultivation, whether ornamental or commercial, and whether the plantations have given rise to self-propagating stands or not. It is evident that this decision was the only possible one considering the present stage of our knowledge and the resources available. It is also clear, on the other hand, that this will lead to still more difficulties in the future, e.g. as the renewal of former natural stands by seed of foreign provenance becomes the rule rather than the exception.

ACKNOWLEDGEMENTS

Our respectful thanks are primarily due to the members of the Committee for Mapping the Flora of Europe, for expert work and help given in many different ways throughout the preparation of the second volume. Here again, we also received valuable information and additional help from a number of specialists in certain taxonomic groups or special fields of systematics and geobotany. For comments and suggestions on the manuscript our thanks are especially due to Dr. Leena Hämet-Ahti (Helsinki).

We highly appreciate the continuing readiness of the Department of Botany, University of Helsinki (Head: Professor Aarno Kalela), to provide working facilities for the Secretariat.

From 1967 onwards, the annual expenditure of the Secretariat was covered by means put at its disposal by the Ministry of Education of Finland.

We are particularly grateful to the Finnish Biological Society Vanamo for undertaking the publication of Atlas Florae Europaeae.

GYMNOSPERMAE

CONIFEROPSIDA

PINACEAE

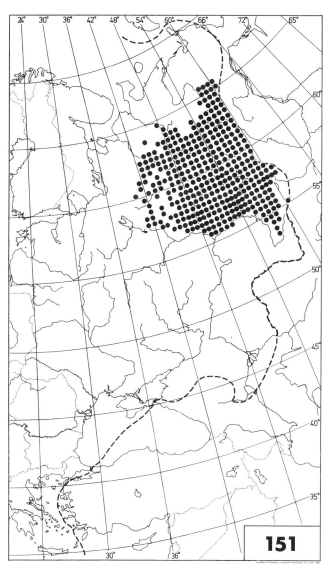

Abies sibirica

Abies sibirica Ledeb. — Map 151.

Total range. Fiziko-geogr. Atlas Mira, p. 242 (Moskva 1964); N. A. Borodina et al., Derev'ja i kustarniki, p. 214 (Moskva 1966).

Abies alba Miller — Map 152.

Abies pectinata (Lam.) DC.

Notes. Rs(C, W) added (not given in Fl. Eur.). Doubtfully native in Normandie, Ga (given as native in Fl. Eur.); e.g. MJW 1965: 393. Recorded from Tu (a single tree), although a confusion with *Abies bornmuelleriana* Mattf. seems likely, and now extinct; see P. H. Davis, Fl. Turkey 1: 70 (Edinburgh 1965), D.A. Webb, Proc. Roy. Irish Acad. 65 B 1: 10 (1966).

Total range. Endemic to Europe.

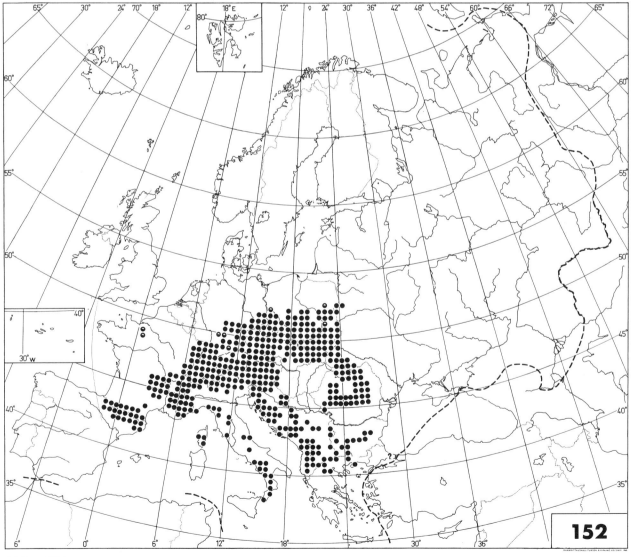

Abies alba

Abies nebrodensis (Lojac.) Mattei — Map 153.

Abies alba subsp. nebrodensis (Lojac.) Nitz.

Taxonomy. According to T. G. Nitzelius, Lust-gården 1968: 178—181 (1969), morphologically intermediate between *Abies alba* and the Algerian *A. numidica* De Lannoy. Nitzelius also points out the occurrence in the Calabrian populations of *A. alba* of individual trees closely related to *A. ne-brodensis.*

Total range. Endemic to Sicilia.

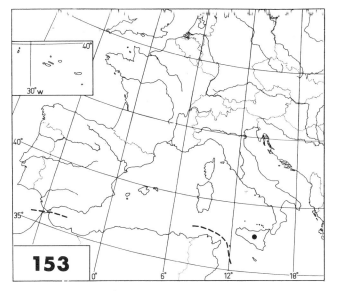

Abies nebrodensis

Abies borisii-regis Mattf. — Map 154.

Taxonomy. More or less intermediate between *Abies alba* and *A. cephalonica,* and most probably derived from introgressive hybridization between these two. T. G. Nitzelius (Lustgården 1968: 164—165, 178 (1969)) does not accord it specific status, but simply divides the material between the supposed parental species.

Notes. Al, Bu, Gr, Ju.

Total range. Endemic to Europe.

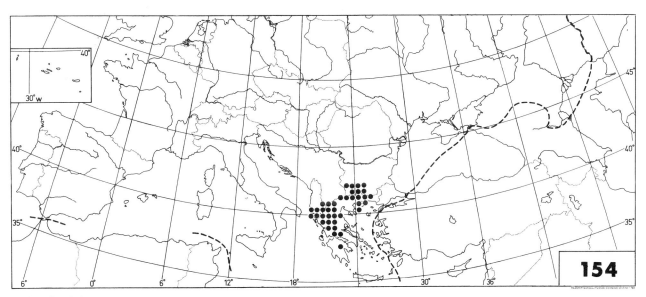

Abies borisii-regis

12

Abies cephalonica Loudon — Map 155.

Total range. Endemic to Europe.

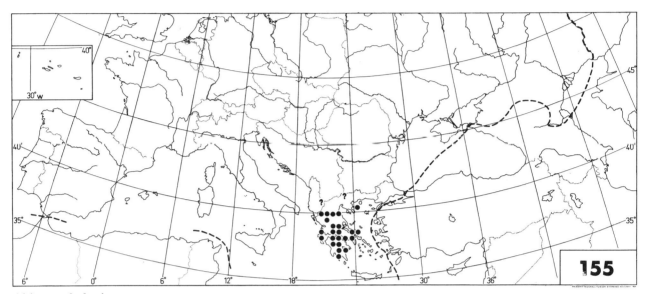

Abies cephalonica

Abies pinsapo Boiss. — Map 156.

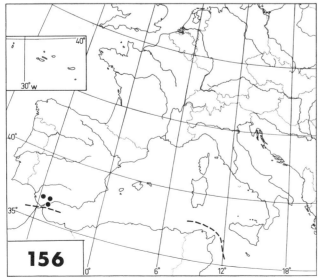

Abies pinsapo

Total range. Endemic to Europe (Fl. Eur.). However, the Moroccan *Abies maroccana* Trabut (*A. pinsapo* subsp. *maroccana* (Trabut) Emberger & Maire) has at times been considered conspecific with the Spanish *A. pinsapo* s. str.; see T. G. Nitzelius, Lustgården 1968: 160—163 (1969). MJW 1965: map 20b; Fenaroli 1967: 25.

Picea abies (L.) Karsten   Map 157.

Taxonomy. The variation pattern of *Picea abies* has proved to be essentially of cline-like character. Consequently, the recognition of taxa at subspecific level does not seem to be very well founded, although the extremes have even been treated as different species, *P. abies* s.str. and *P. obovata* Ledeb. *P. fennica* (Regel) Komarov covers some of the intermediates, and is considered by E. G. Bobrov, Nov. Syst. Pl. Vasc. (Leningrad) 7: 31 (1971), to have resulted from introgressive hybridization between the two extremes; see also P. Kallio et al., Ann. Univ. Turkuensis A II 47 (Rep. Kevo Subarctic Res. Stat. 8): 80—82 (1971). A considerable part of the records for *P. fennica* have probably been included in the map of subsp. *abies*, and similar variants present in the C. European mountains are sometimes treated as *P. abies* subsp. *alpestris* (Brügger) Domin.

Notes. Because of the difficulties in making a clear-cut and concise division of the records between the two subspecies, an additional collective map is presented. Gr and Rs(W) added (not given in Fl. Eur.), Hu added ([Hu] in Fl. Eur.).

Total range. MJW 1965: map 20d.

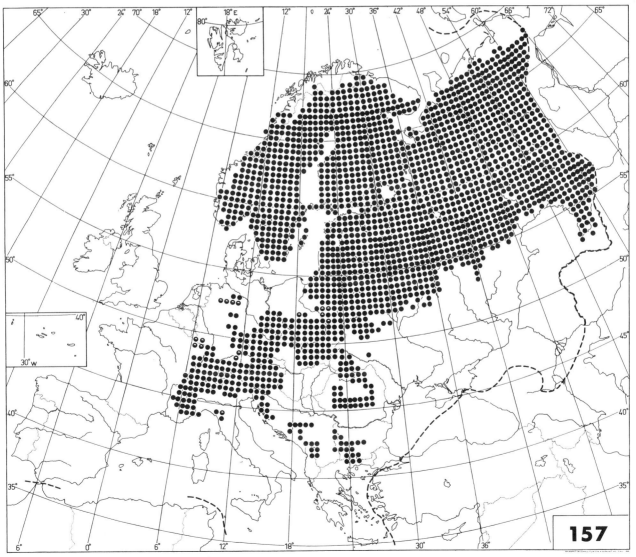

Picea abies coll.

14

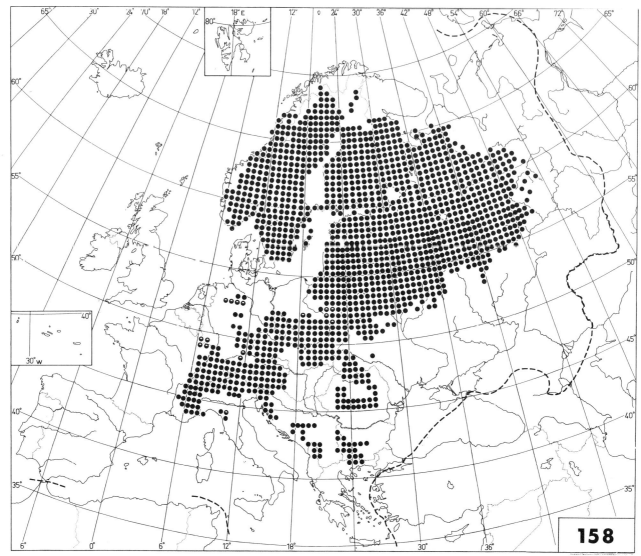

Picea abies subsp. **abies**

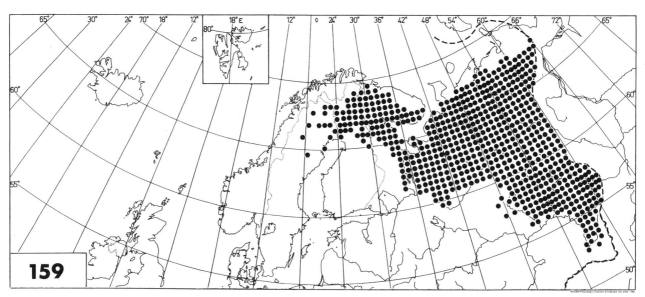

Picea abies subsp. **obovata**

P. abies subsp. **abies** — Map 158.

Picea excelsa (Lam.) Link; P. vulgaris Link; Incl. P. abies subsp. alpestris (Brügger) Domin and P. fennica (Regel) Komarov, ? pro parte.

Taxonomy. See above. Also V. Parfenov, Nov. Syst. Pl. Vasc. (Leningrad) 8: 4—11 (1971).

Notes. Gr and Rs (N, W) added (not given in Fl. Eur.), Hu added ([Hu] in Fl. Eur.).

Total range. Endemic to Europe. A map showing the total range of *P. fennica* in E. G. Bobrov, Nov. Syst. Pl. Vasc. (Leningrad) 7: 17 (1971).

P. abies subsp. **obovata** (Ledeb.) Domin — Map 159.

Picea obovata Ledeb.

Taxonomy. See under the species.

Nomenclature. J. Holub, Preslia 38: 79 (1966).

Notes. The dot in No (Sør-Varanger) is according to O. Dahl, Nyt Mag. Naturvid. 69: 237 (1934). Probably more common in northern parts of Su than indicated. Also recorded from the Alps, although not mentioned in Fl. Eur., and not mapped here; see G. Priehäusser, Ber. Bayer. Bot. Ges. 35: 96—104 (1962). In addition, the *obovata* cone scale type is locally present at least in Po (H. Chylarecki & M. Giertych, Arbor. Kórnickie 14: 39—71 (1969)).

Total range. MJW 1965: map 20d; N. A. Borodina et al., Derev'ja i kustarniki, p. 220 (Moskva 1966).

Picea omorika (Pančić) Purkyně — Map 160.

Total range. Endemic to Europe.

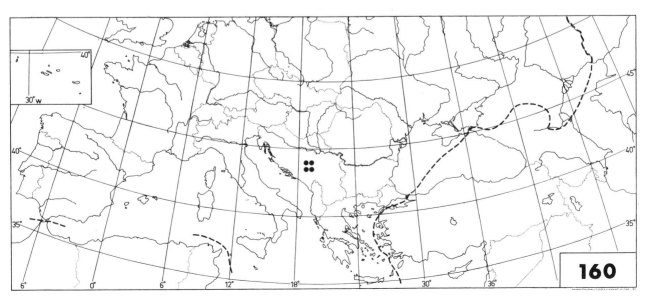

Picea omorika

Pinus nigra Arnold

Pinus maritima Miller, nom. ambig.

Taxonomy and nomenclature. O. Schwarz, Notizbl. Bot. Garten Berlin 13: 226—243 (1938), 14: 135—136 and 381—384 (1939); E. Janchen & H. Neumayer, Österr. Bot. Zeitschr. 91: 215 (1942); E. Huguet del Villar, Ber. Schweiz. Bot. Ges. 57: 149—155 (1947); P. Fukarek. Rad. Polj. Šum. Fakult. Sarajevu 3: 3—92 (1958); A. Ž. Lovrić, Österr. Bot. Zeitschr. 119: 569—570 (1971).

Notes. Cr omitted (given in Fl. Eur.), Ga added (not given in Fl. Eur.).

Total range. MJW 1965: map 22b; W. B. Critchfield & E. L. Little, U.S. Dept. Agricult. Forest Service Misc. Publ. 991: map 27 (1966); Fenaroli 1967: 42.

P. nigra subsp. nigra — Map 164.

Pinus austriaca Höss; P. nigricans Host.

Notes. Al, Au, Gr, It, Ju. Some of the southeasternmost dots refer to *Pinus nigra* coll.
Total range. Endemic to Europe.

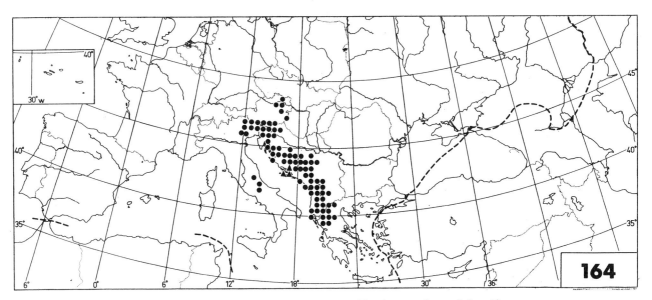

● = **Pinus nigra** subsp. **nigra**   ▲ = **P. nigra** subsp. **dalmatica**

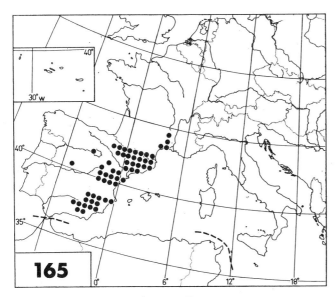

Pinus nigra subsp. **salzmannii**

P. nigra subsp. salzmannii (Dunal) Franco — Map 165.

Pinus clusiana Clemente; P. pyrenaica auct., an Lapeyr., nom. dub.; P. salzmannii Dunal

Taxonomy and nomenclature. The closely related *Pinus nigra* subsp. *mauretanica* (Maire & Peyerimhoff) Heywood has also been reported from Spain, although not mentioned in Fl. Eur.; see V. H. Heywood & P. W. Ball, Feddes Repert. 66: 150 (1962).
Notes. Ga, Hs.
Total range. MJW 1965: map 22b (including subsp. *mauretanica*); Fenaroli 1967: 42.

P. nigra subsp. **laricio** (Poiret) Maire — Map 166,

Pinus laricio Poiret, non Santi; Incl. P. calabrica Delamare

Notes. Co, It, Sa.
Total range. Endemic to Europe.

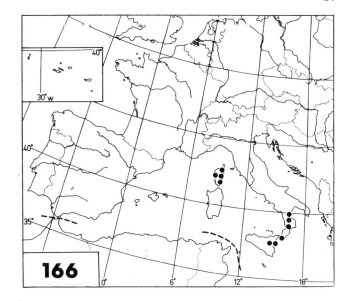

Pinus nigra subsp. **laricio**

P. nigra subsp. **dalmatica** (Vis.) Franco — Map 164.

Taxonomy. M. Vidakovič, Glasnik Šumske Pokuse (Zagreb) 13: 183 — 193, 244—245 (1957).
Notes. Ju. Mapped according to R. Domac, Ber. Geobot. Inst. Rübel (Zürich) 36: 104 (1965).
Total range. Endemic to Europe.

P. nigra subsp. **pallasiana** (Lamb.) Holmboe — Map 167.

Pinus pallasiana Lamb.; Incl. P. banatica (Georgescu & Ionescu) Georgescu & Ionescu

Notes. Bu, Gr, Ju, Rm, Rs(K), Tu.
Total range. MJW 1965: map 22b; Fenaroli 1967: 42.

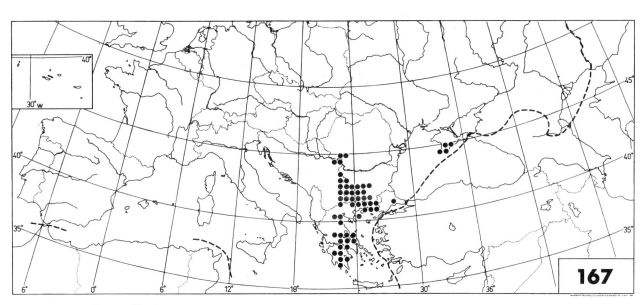

Pinus nigra subsp. **pallasiana**

Pinus sylvestris L. — Map 168.

Incl. Pinus kochiana Klotzsch ex Koch; P. fominii Kondrat.; P. hamata (Steven) D. Sosn.

Taxonomy. Extremely polymorphic, see the list of varieties in Fl. Eur., and e.g. L. F. Pravdin, Sosna obyknovennaja (Moskva 1964; English translation Jerusalem 1969). Mapped as one collective taxon.

Notes. In Da extinct as a native in the eighteenth century, but later on frequently self-sown from planted stands ([Da] in Fl. Eur.). In Ho, some native post-glacial stands possibly survived long enough to give rise to part of the recent occurrences (*Ho in Fl. Eur.), the remainder of which have originated from cultivation during the last few centuries. In Lu doubtfully native and/or extinct (given as native in Fl. Eur.). Rs(E) added (not given in Fl. Eur.).

Total range. MJW 1965: map 21d; N. A. Borodina et al., Derev'ja i kustarniki, p. 236 (Moskva 1966); W. B. Critchfield & E. L. Little, U.S. Dept. Agricult. Forest Service Misc. Publ. 991: map 32 (1966); Fenaroli 1967: 33; N. T. Mirov, The Genus Pinus, pp. 276—277 (New York 1967); J. Holub & V. Jirásek, Folia Geobot. Phytotax. (Praha) 3: 319 (1968).

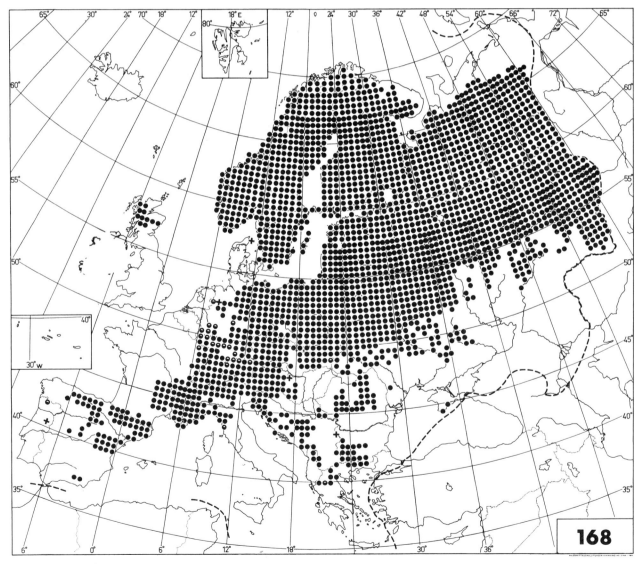

Pinus sylvestris

Pinus mugo Turra — Map 169.

Pinus montana Miller; P. mughus Scop.; Incl. P. pumilio Haenke

Notes. Gr omitted (given in Fl. Eur.).
Total range. Endemic to Europe.

Pinus uncinata Miller ex Mirbel — Map 170.

Taxonomy. Pinus rotundata Link and *P. uliginosa* Neumann have been provisionally mapped as a separate entity, although there is no full agreement regarding its delimitation from *P. uncinata* s.str. The populations of the C. and E. Alps, between 8° and 13° E (*P. rostrata* auct. alp. orient.), also seem to differ from *P. uncinata* of the Pyrenees and the SW. Alps.

Notes. P. uncinata s.str.: Au added (not given in Fl. Eur.). *P. rotundata* + *P. uliginosa:* Au, Cz, Ge, Po.
Total range. Endemic to Europe.

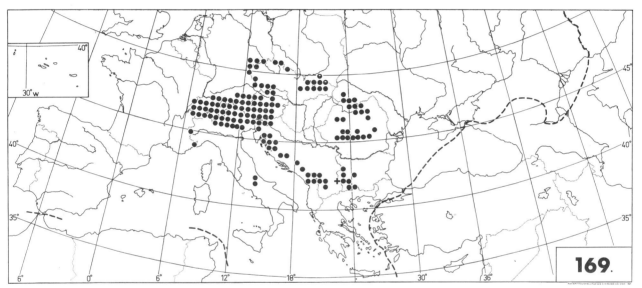

Pinus mugo

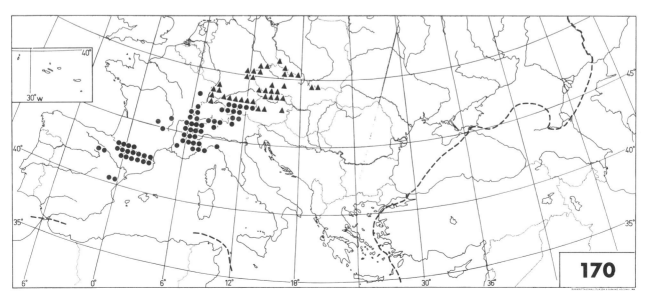

Pinus uncinata   ● = P. uncinata sensu stricto   ▲ = P. rotundata + P. uliginosa

Pinus heldreichii Christ — Map 171.

Incl. Pinus leucodermis Antoine

Taxonomy. In accordance with the treatment of F. Markgraf, Mitt. Deutsch. Dendrol. Ges. 1931: 250—255, J. Schultze-Motel, Die Kulturpflanze Beih. 4: 41 (1966), P. Fukarek, Bot. Jahrb. 86: 449 (1967), and others, *P. leucodermis* is considered a variant of *P. heldreichii*, and is not mapped separately.

Total range. Endemic to Europe.

Pinus halepensis Miller — Map 172.

Notes. Al added (not given in Fl. Eur.), not native in Cr (? Cr in Fl. Eur.).

Total range. W. B. Critchfield & E. L. Little, U.S. Dept. Agricult. Forest Service Misc. Publ. 991: map 31 (1966); N.T. Mirov, The Genus Pinus, p. 251 (New York 1967); Fenaroli 1967: 46; J. Holub & V. Jirásek, Folia Geobot. Phytotax. (Praha) 3: 310 (1968). The two last-mentioned maps are erroneous in several details, partly owing to confusion with *P. brutia.*

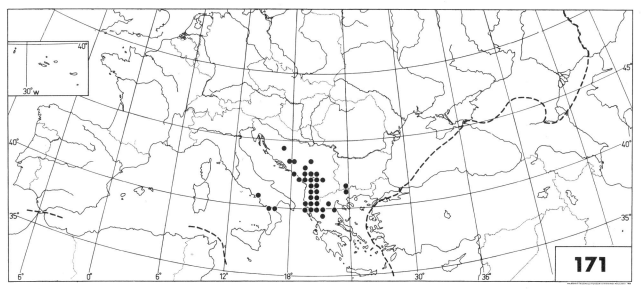

Pinus heldreichii

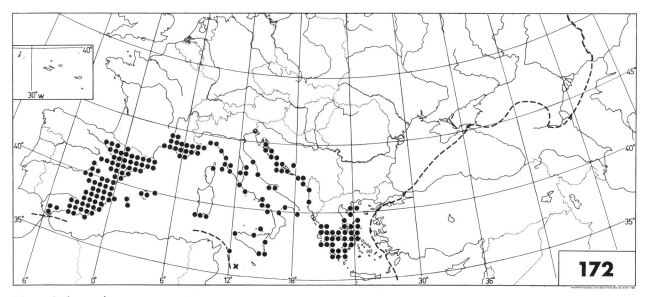

Pinus halepensis

Pinus brutia Ten. — Map 173.

Incl. Pinus pityusa Steven and P. stankewiczii (Suk.) Fomin

Taxonomy. For the chemotaxonomical relationships between *Pinus brutia* and the two taxa here included in it, see N.T. Mirov et al., Phytochemistry 5: 97—102 (1966), and N. T. Mirov, The Genus Pinus, pp. 552—553 (New York 1967).

Notes. Gr added (not given in Fl. Eur.), doubtfully native in It (given as native in Fl. Eur.).

Total range. W. B. Critchfield & E. L. Little, U.S. Dept. Agricult. Forest Service Misc. Publ. 991: map 31 (1966); N. T. Mirov, The Genus Pinus, p. 251 (New York 1967).

Pinus pinea L. — Map 174.

Notes. Cr omitted (given in Fl. Eur.), in Tu doubtfully native (given as native in Fl. Eur.). There are conflicting opinions concerning the status, whether native or introduced, in, e.g., Gr, It and Ju.

Total range. W. B. Critchfield & E. L. Little, U.S. Dept. Agricult. Forest Service Misc. Publ. 991: map 25 (1966); Fenaroli 1967: 52 (1967); N. T. Mirov, The Genus Pinus, p. 240 (New York 1967).

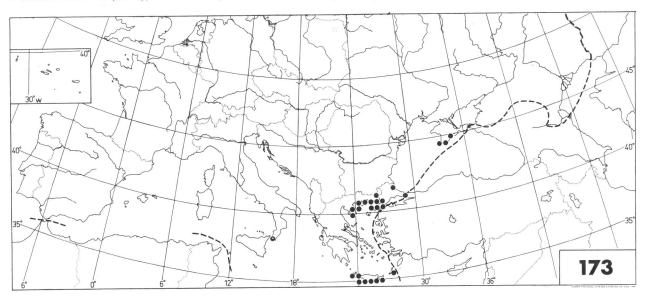

Pinus brutia

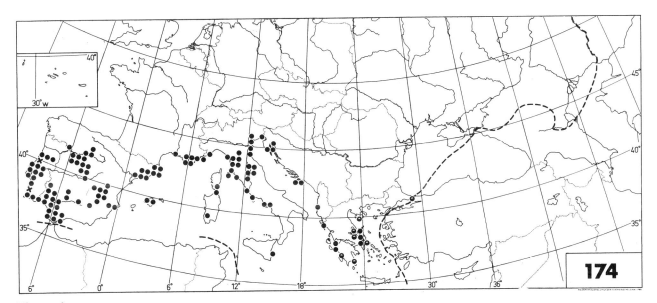

Pinus pinea

Pinus cembra L. — Map 175.

Notes. Ju omitted (given in Fl. Eur.).
Total range. Endemic to Europe.

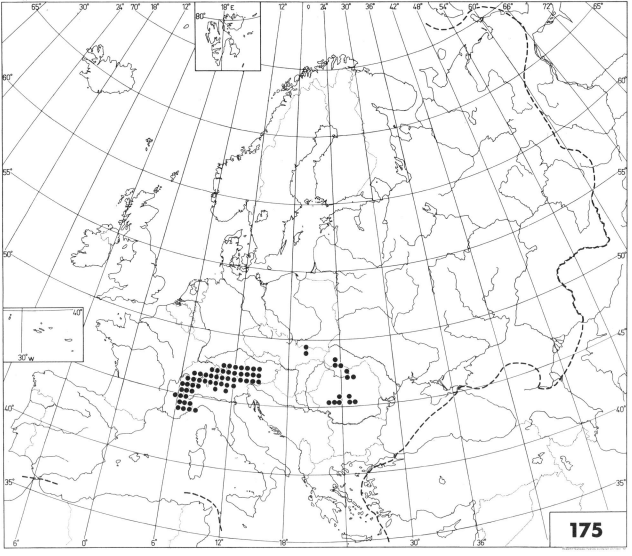

Pinus cembra

Pinus sibirica Du Tour — Map 176.

Pinus cembra subsp. sibirica (Du Tour) Krylov

Total range. Fiziko-geogr. Atlas Mira, p. 92 and 112 (Moskva 1964); MJW 1965: map 22c; N. A. Borodina et al., Derev'ja i kustarniki, p. 230 (Moskva 1966); W. B. Critchfield & E. L. Little, U.S. Dept. Agricult. Forest Service Misc. Publ. 991: map 4 (1966); Y. de Ferré, Bull. Soc. Hist. Nat. Toulouse 102: 355 (1966); N. T. Mirov, The Genus Pinus, p. 264—265 (New York 1967); A. K. Skvortsov, Proc. Study Fauna Fl. USSR, N. S. (Bot.) 15: 86 (1968).

Pinus peuce Griseb. — Map 177.

Notes. Gr added (?Gr in Fl. Eur.).
Total range. Endemic to Europe.

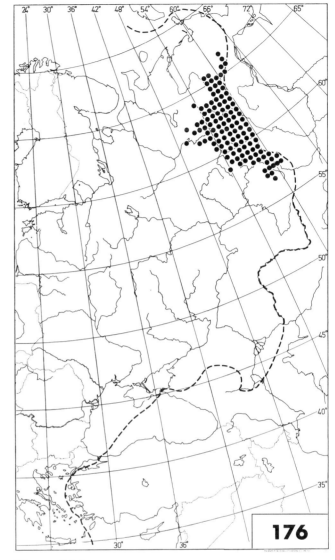

Pinus sibirica

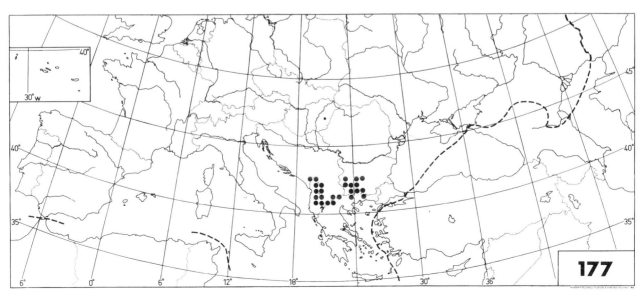

Pinus peuce

CUPRESSACEAE

Cupressus sempervirens L. — Map 178.

 Notes. Most probably not native in Gr (given as native in Fl. Eur.), or along the Adriatic coast.
 Total range. Fenaroli 1967: 60; G. Krüssmann, Die Bäume Europas, p. 95 (Berlin & Hamburg 1968).

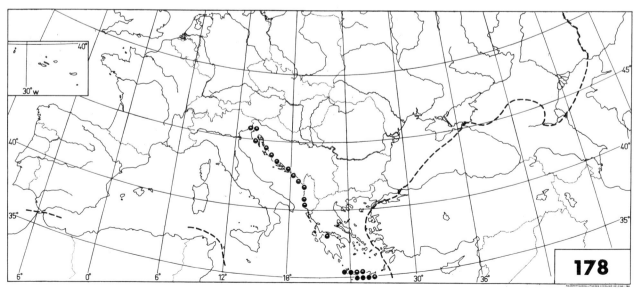

Cupressus sempervirens

Tetraclinis articulata (Vahl) Masters   Map 179.

Total range. M. Rikli, Das Pflanzenkleid der Mittelmeerländer I: 220 (Bern 1943); Fiziko-geogr. Atlas Mira, p. 132 (Moskva 1964).

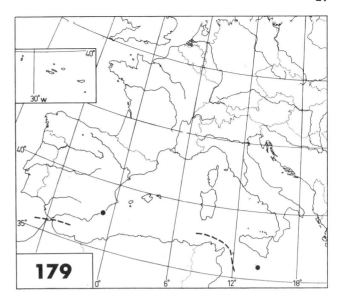

Tetraclinis articulata

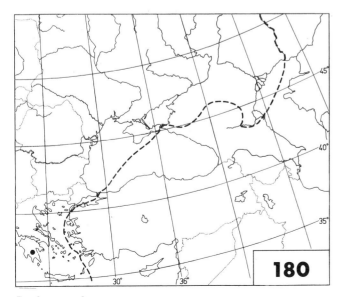

Juniperus drupacea Labill. — Map 180.

Arceuthos drupacea (Labill.) Antoine & Kotschy

Total range. Th. Schmucker, Silvae Orbis, map 59 (Berlin 1942).

Juniperus drupacea

Juniperus communis L.   Map 181.

Taxonomy. J. do Amaral Franco, Bol. Soc. Brot. 36: 101—120 (1962), H. J. Welch, Gard. Chron. 166 (5): 14 —15 (1969). Considering the great phenotypic plasticity of the species, and the lack of definite biosystematic data, it seems somewhat premature to recognize the morphological extremes as subspecies. Consequently, a collective map has been included.

Total range. Hultén CP 1964: map 66; MJW 1965: map 22d.

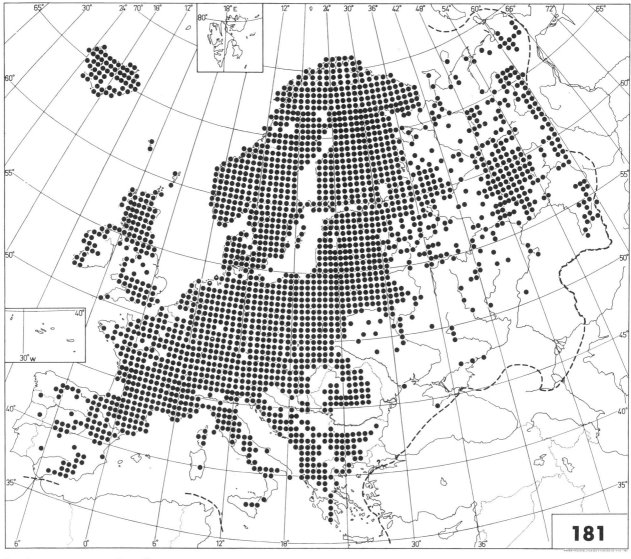

Juniperus communis coll.

J. communis subsp. communis — Map 182

Notes. Subsp. *alpina* and subsp. *hemisphaerica* have in several cases not been kept strictly apart from subsp. *communis*. Consequently, the map is more or less tentative, especially in overlapping areas, and includes at least a number of intermediate plants.

Total range. Hultén CP 1964: map 66.

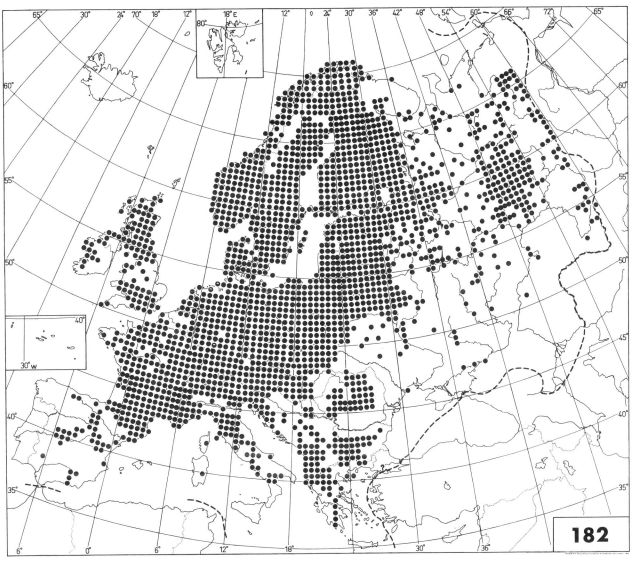

Juniperus communis subsp. **communis**

J. communis subsp. **hemisphaerica** (J. & C. Presl) Nyman — Map 183.

Juniperus depressa Steven

Nomenclature. D. R. Hunt & H.J. Welch, Taxon 17: 545 (1968).
Notes. The map is provisional.
Total range. Hultén CP 1964: map 66.

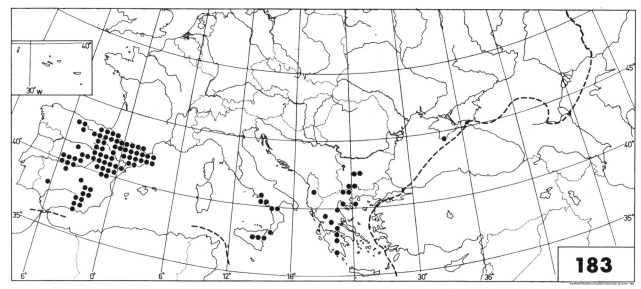

Juniperus communis subsp. **hemisphaerica**

J. communis subsp. **alpina** (Neilr.) Čelak. — Map 184.

Juniperus communis var. montana Aiton; J. communis subsp. nana Syme; J. nana Willd.; J. sibirica Burgsd.

Taxonomy and nomenclature. M. Breistroffer, Bull. Soc. Bot. Fr. 110 (sess. extraord.): 50—51 (1963); J. Holub, Preslia 38: 79 (1966); K. Browicz et al., Atlas Distrib. Trees Shrubs Poland 10: 11—14 (1971). It may be questioned whether all the material mapped really belongs together taxonomically; see G. Turesson, Bot. Not. 114: 440—442 (1961), P. H. Davis, Fl. Turkey I: 79 (1965).

Total range. Hultén CP 1964: map 66 (as var. *montana* Aiton); K. Browicz et al., Atlas Distrib. Trees Shrubs Poland 10: map 11 (1971).

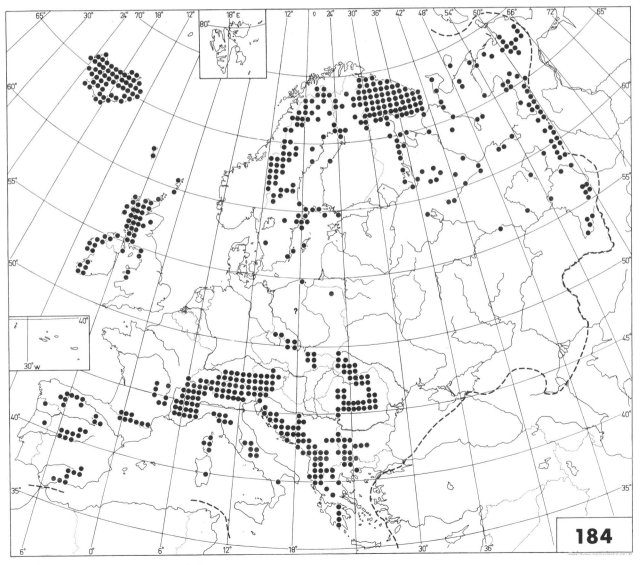

Juniperus communis subsp. **alpina**

Juniperus oxycedrus L.

Taxonomy. J. do Amaral Franco, Feddes Repert. 68 (Fl. Eur. Not. Syst. 2): 163—167 (1963).

Notes. Bu and Cr added (not given in Fl. Eur.).

Total range. J. do Amaral Franco, Feddes Repert. 68: folded map between pages 164 and 165 (1963) (the map not present in Fl. Eur. Not. Syst. 2 reprints); G. Krüssmann, Die Bäume Europas, p. 101 (Berlin & Hamburg 1968).

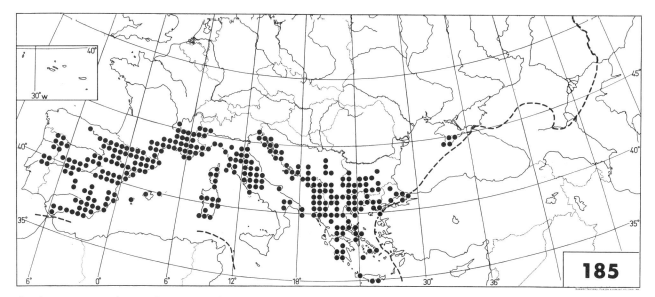

Juniperus oxycedrus subsp. **oxycedrus**

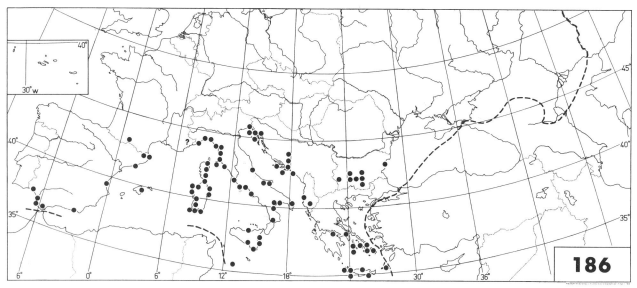

Juniperus oxycedrus subsp. **macrocarpa**

J. oxycedrus subsp. **oxycedrus** — Map 185.

Juniperus rufescens Link

Notes. The map is partly collective, owing to confusion with subsp. *macrocarpa*.

J. oxycedrus subsp. **macrocarpa** (Sibth. & Sm.) Ball — Map 186.

Juniperus macrocarpa Sibth. & Sm.; J. umbilicata Godron

Notes. Bl, Co, Gr, Hs, It, Sa, Si (J. do Amaral Franco, Feddes Repert. 68: 166 (1963)). Al, Bu, Cr, Ga and Ju added.

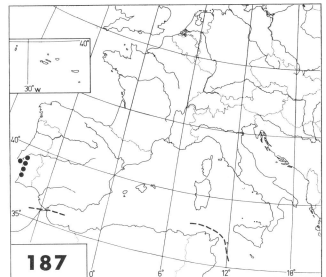

J. oxycedrus subsp. **transtagana** Franco — Map 187.

Juniperus oxycedrus subsp. rufescens auct. lusit., non Deb.

Total range. Endemic to Europe.

Juniperus oxycedrus subsp. **transtagana**

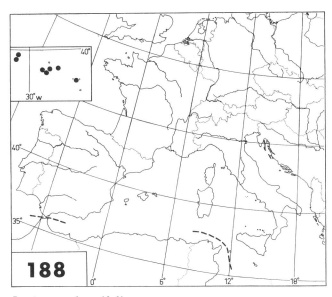

Juniperus brevifolia

Juniperus brevifolia (Seub.) Antoine — Map 188.

Total range. Endemic to the Açores.

Juniperus phoenicea L. — Map 189.

Total range. G. Krüssmann, Die Bäume Europas, p. 101 (Berlin & Hamburg 1968).

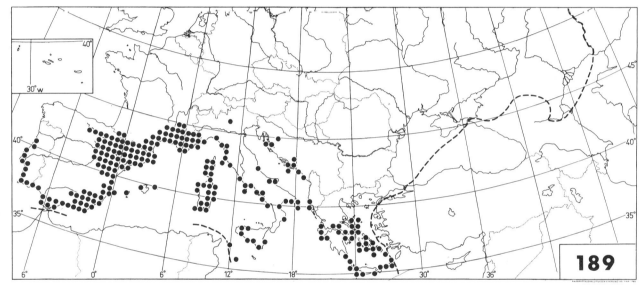

Juniperus phoenicea

Juniperus thurifera L. — Map 190.

Notes. Co added (not given in Fl. Eur.).

Total range. M. Rikli, Das Pflanzenkleid der Mittelmeerländer, p. 418 (Bern 1943); G. Krüssmann, Die Bäume Europas, p. 201 (Berlin & Hamburg 1968).

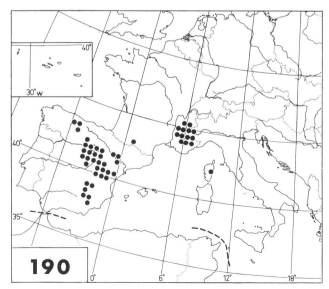

Juniperus thurifera

Juniperus foetidissima Willd. — Map 191.

Total range. S. J. Sokolov, Derev'ja i kustarniki SSSR I: 339 (1949).

Juniperus excelsa Bieb. — Map 192.

Notes. Al added (not given in Fl. Eur.), Cr omitted (given in Fl. Eur.).

Total range. M. Rikli, Das Pflanzenkleid der Mittelmeerländer, p. 419 (Bern 1943); S. J. Sokolov, Derev'ja i kustarniki SSSR I: 359 (1949); Fiziko-geogr. Atlas Mira, p. 92 (Moskva 1964).

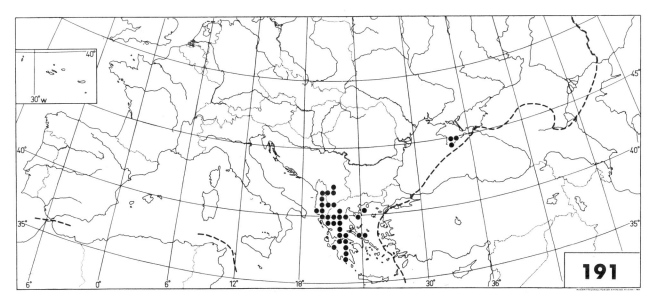

Juniperus foetidissima

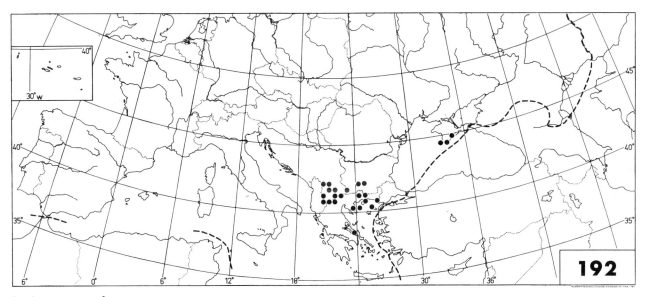

Juniperus excelsa

Juniperus sabina L. — Map 193.

Notes. Al, Gr and Rs(W) added (not given in Fl. Eur.), Rs(C) confirmed (Rs(?C) in Fl. Eur.).

Total range. MJW 1965: map 23a; K. Browicz & M. Gostyńska-Jakuszewska, Atlas Distrib. Trees Shrubs Poland 7: map 12 (1968).

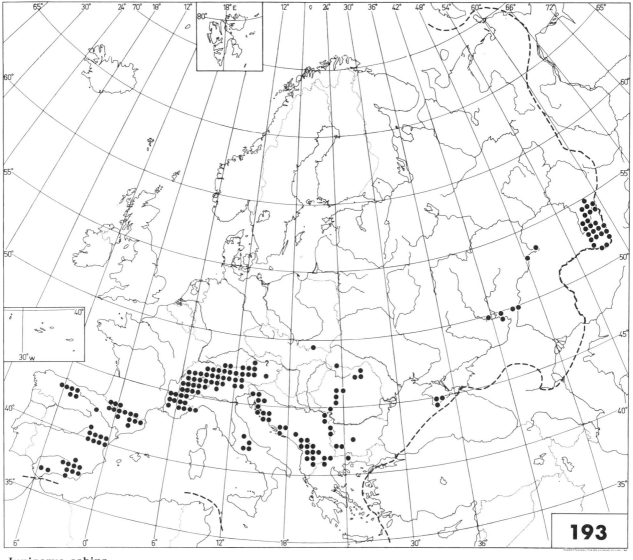

Juniperus sabina

TAXOPSIDA

TAXACEAE

Taxus baccata L. — Map 194.

Notes. In Ho doubtfully native (given as native in Fl. Eur.).

Total range. MJW 1965: map 19d; Fenaroli 1967: 64; E. Jäger, Feddes Repert. 79: 178 (1968) (the genus);
K. Browicz & M. Gostyńska-Jakuszewska, Atlas Distrib. Trees Shrubs Poland 8: map 1 (1969).

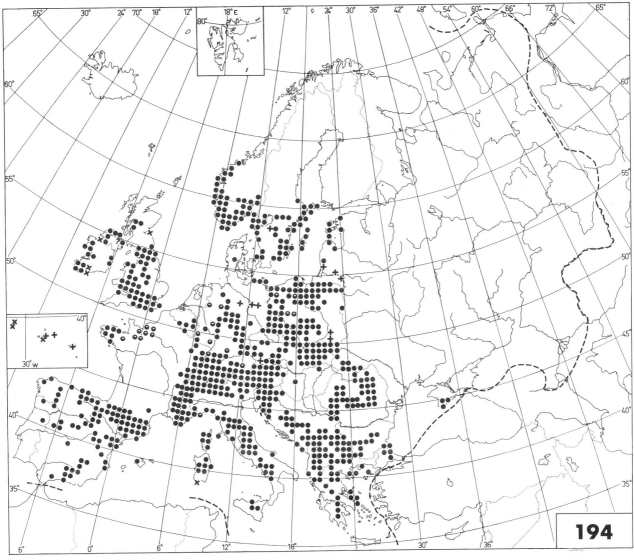

Taxus baccata

GNETOPSIDA

EPHEDRACEAE

Ephedra fragilis Desf.

Ephedra altissima sensu Willk.

Notes. It omitted (given in Fl. Eur.).
Total range. M. Welten, Ber. Schweiz. Bot. Ges. 67: 38 (1957).

E. fragilis subsp. fragilis — Map 195.

Notes. Bl, Hs, Lu, Si. H. Riedl, Scient. Pharmac. 35: 227 (1967).

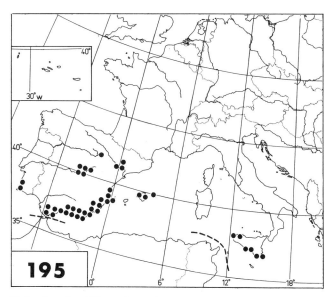

195

Ephedra fragilis subsp. **fragilis**

E. fragilis subsp. campylopoda (C. A. Meyer) K. Richter — Map 196.

Notes. Al, Bu, Cr, Gr, Ju, Tu.
Total range. M. Welten, Ber. Schweiz. Bot. Ges. 67: 38 (1957).

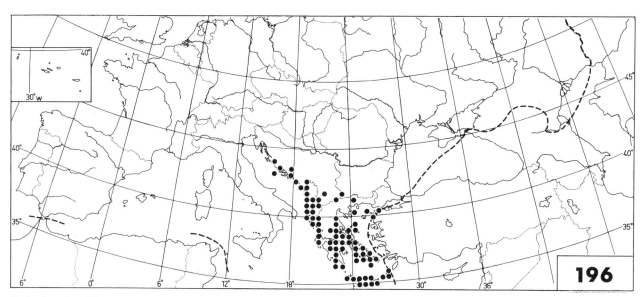

196

Ephedra fragilis subsp. **campylopoda**

Ephedra distachya L.

Ephedra vulgaris L.C.M. Richard

Notes. Rs(K) added (not given in Fl. Eur.).

Total range. MJW 1965: map 19c.

E. distachya subsp. **distachya** — Map 198.

Taxonomy. The plant in eastern Europe and farther east has recently been referred to *Ephedra distachya* subsp. *monostachya* (L.) H. Riedl, Scient. Pharmac. 35: 228 (1967).

Notes. Al, Bu, Co, Cz, Ga, Gr, Hs, Hu, It, Ju, Rm, Rs(C, W, K, E), Sa, Si.

E. distachya subsp. **helvetica** (C. A. Meyer) Ascherson & Graebner — Map 197.

Ephedra helvetica C. A. Meyer

Taxonomy. Considered a distinct species by H. Riedl, Scient. Pharmac. 35: 227 (1967).

Notes. Ga, He, It. The easternmost dot (Trento) according to H. Riedl, Scient. Pharmac. 35: 227 (1967).

Total range. Endemic to Europe.

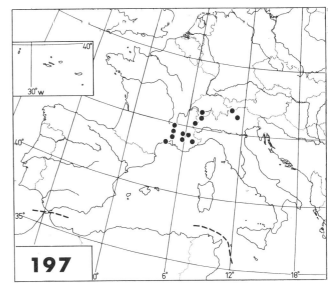

Ephedra distachya subsp. **helvetica**

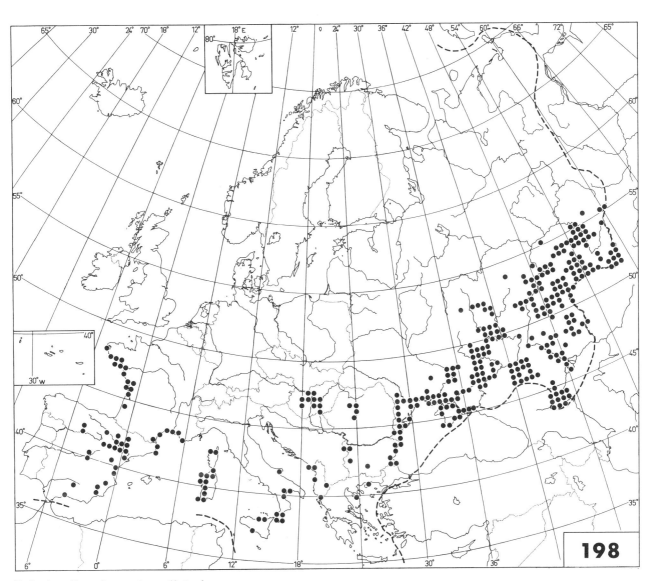

Ephedra distachya subsp. **distachya**

Ephedra major Host

Ephedra nebrodensis Tineo ex Guss.; E. scoparia Lange

Notes. Al and Tu added (not given in Fl. Eur.).

E. major subsp. **major** — Map 199.

Notes. Al, Ga, Gr, Hs, It, Ju, Sa, Si, Tu.

E. major subsp. **procera** (Fischer & C. A. Meyer) Markgraf — Map 200.

Ephedra procera Fischer & C. A. Meyer

Taxonomy. P. H. Davis, Fl. Turkey I: 85 (1965). Considered by H. Riedl, Scient. Pharmac. 35: 228 (1967), to be a minor eastern variant of *E. major*, not deserving subspecific status.

Notes. Ju added (not given in Fl. Eur.). See H. Riedl, Scient. Pharmac. 35: 228 (1967).

Total range. Given as endemic to Europe in Fl. Eur. However, the taxon was described on the basis of material from Caucasia, and is rather widely distributed in the Near East.

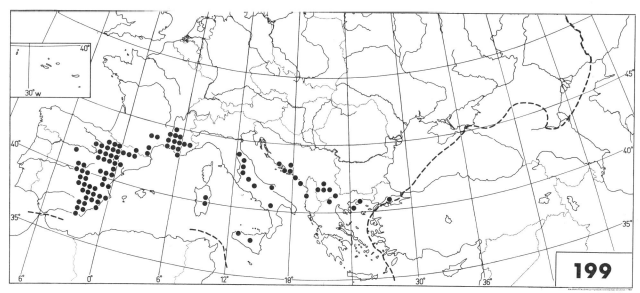

Ephedra major subsp. **major**

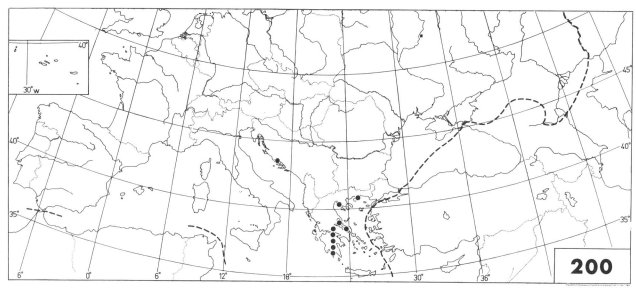

Ephedra major subsp. **procera**

INDEX TO VOL. 1.

Names appearing in the text or as synonyms, and their respective page numbers, are given *in italics*.

INDEX TO VOL. 2

Names appearing in the text or as synonyms, and their respective page numbers, are given *in italics*.

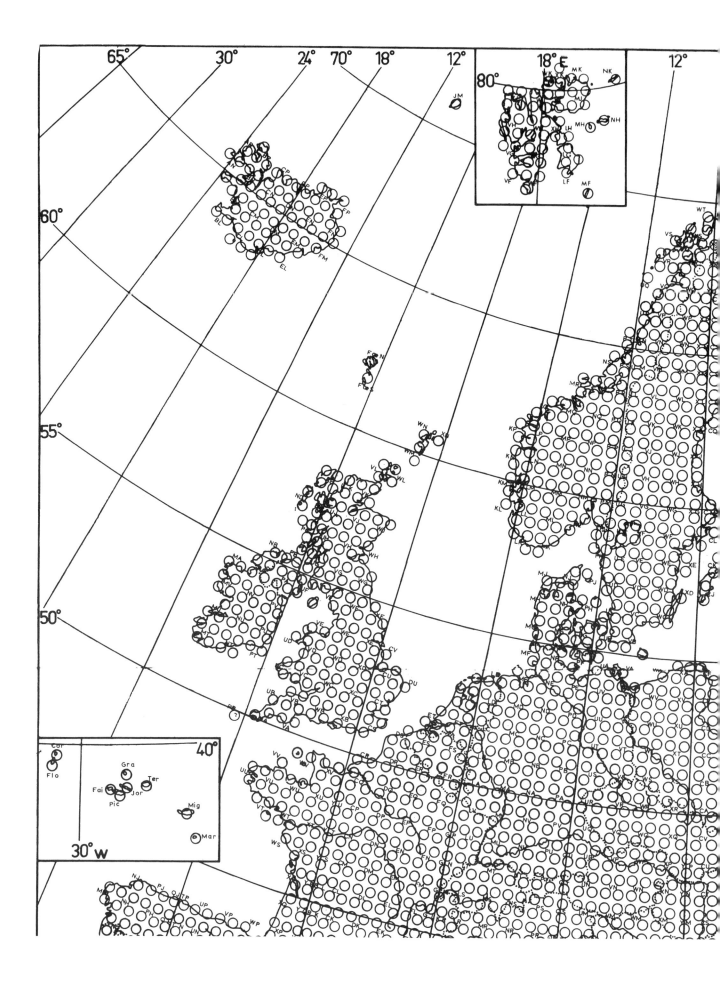

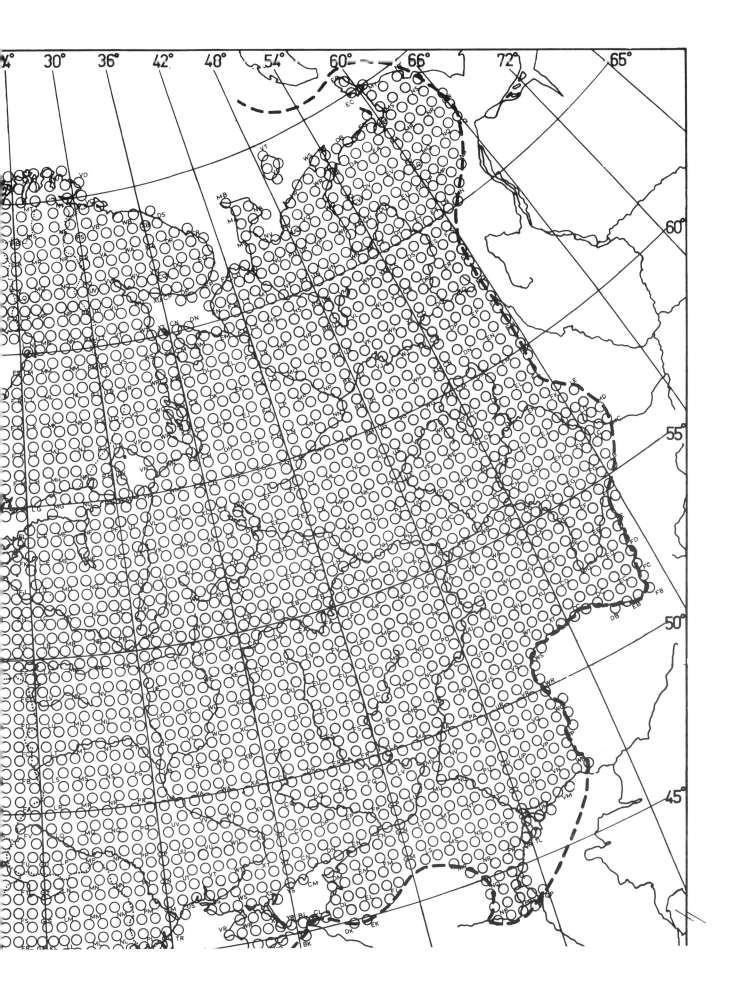

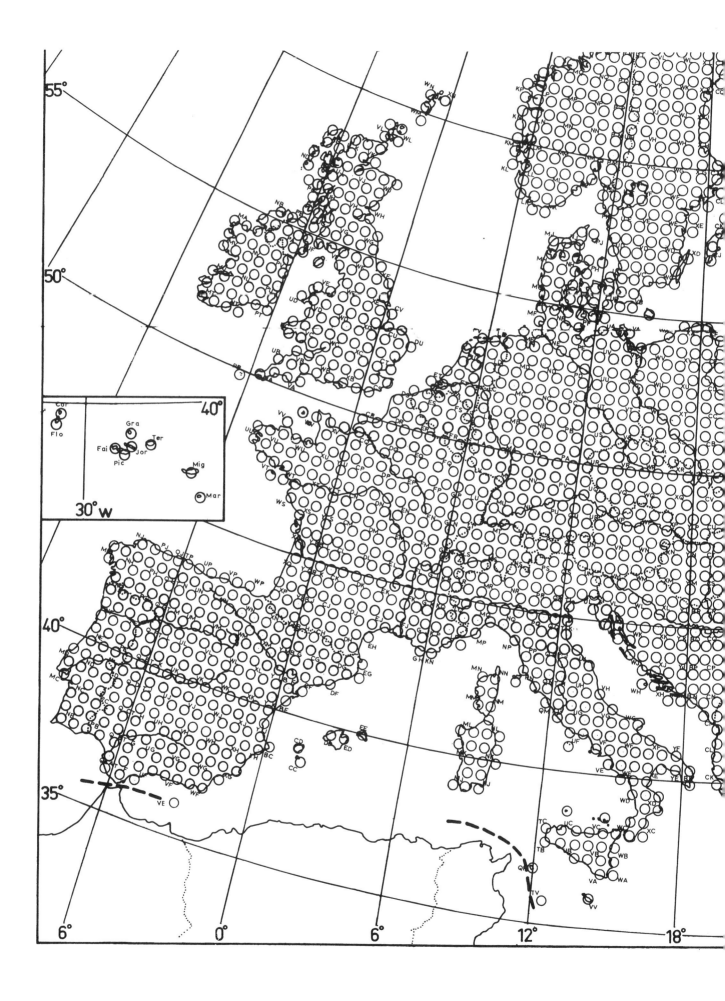

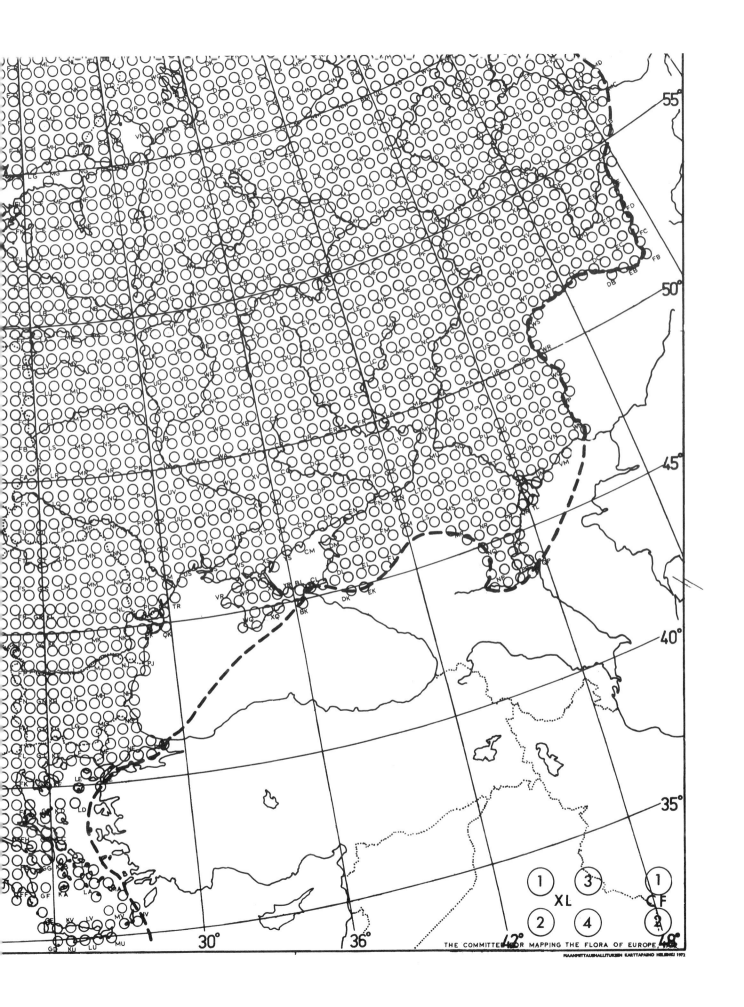

MAANMITTAUSHALLITUKSEN KARTTAPAINO HELSINKI 1972